Implementation of IEC/IEEE 82079-1 Ed. 2

Considering other Sector-Specific Standards

Bibliographic information of Deutsche Nationalbibliothek (German National Library)
The German National Library lists this publication in the German National Bibliography; detailed bibliographic data is available on the internet at https://portal.dnb.de/opac.htm.

tcworld GmbH, Heilbronner Str. 86, 70191 Stuttgart, Germany
Phone +49 711 65704-0
e-mail info@tekom.org
www.tekom.de

ISBN 978-3-96393-004-1 softcover (with IEC/IEEE 82079-1 Ed. 2 Standard)
ISBN 978-3-96393-005-8 e-book PDF
ISBN 978-3-96393-088-1 bundle (softcover and e-book PDF)

Copy-editing: Alexander Kurz, Redaktionsbüro Kurz, Stuttgart, Germany
Layout: Elisabeth Gräfe, www.fraugraefe.de, Germany
Print: Libri Plureos GmbH, Friedensallee 273, 22763 Hamburg, Germany

Implementation of IEC/IEEE 82079-1 Ed. 2

Considering other Sector-Specific Standards

Authors
Martin Tillmann
Michael Fritz / Claudia Klumpp / Martin Rieder /
Roland Schmeling / Stephan Schneider

Contents

1 Introduction

IEC/IEEE 82079-1 Edition 2 Preparation of information for use (instructions for use) of products – Part 1: Principles and general requirements since its publication in 2012 has developed into the most important standard in the field of technical communication. As a so called "horizontal standard" it is applicable to any type of product and all branches of the industry.

Note:
The present commentary always refers to edition 2 of the standard without specifically saying so in each instance.

1.1 Why was the standard revised?

Generally, all standards are subject to a five-year scrutiny and revision cycle. The revision of IEC 82079-1 was initiated as early as 2014. The so called "review report" laid down an outline for this revision: Not only revision errors, such as faulty references, should be corrected; rather, it was clear from the outset that this would be a comprehensive revision dealing in depth with the issues not considered in the first edition; it would contain substantial improvements, new content and address new findings from current practice.

The structure of the standard was identified as a major weakness. Many topics and their related aspects were scattered across the entire standard, resulting in lack of clarity and user-friendliness. Additionally, the comprehensibility of the standard needed improvement, as some passages were incomprehensible or ambiguous.

With regard to the standard's content, the review report mentioned some new topics to which the revised edition should attribute greater importance. In particular, this includes the quality assurance process, which should be given more room–including a definition of pertinent quality criteria. Primarily, the information development process should be represented in greater detail. Also, information on the general principles of and criteria for structuring information were to be added. The structure of step-by-step instructions was to be a key aspect.

1.2 Who shared in the revision?

The revision was handled by a joint working group JWG 16 (Joint Working Group 16) of the two standardisation organisations ISO (International Organization for Standardization) and IEC (International Electrotechnical Commission). The basis for this cooperation is an agreement between the two standardisation organisations to the effect that a standard shall bear only the name of the organisation that took the lead in its preparation. Since IEC always spearheaded the development of 82079-1, the previous standard bore the designation IEC 82079-1 although it was and still is also supported by ISO. It is important to note this, because the acceptance of a particular standard in some industry branches may depend on the standardisation organisation that has published it. For instance, standardisation for mechanical engineering is mainly issued by ISO, making it harder for IEC standards to gain the necessary acceptance by mechanical engineering companies.

Additionally, a third large standardisation organisation shared in preparing the standard, IEEE (Institute of Electrical and Electronics Engineers, see figure 1). Since only IEC and IEEE have an agreement concerning joint publication of standards, the new edition's designation only reads IEC/IEEE. We consider this unfortunate, because ISO is still supporting its publication. Visibility and significance of the standard in the US market, however, will have grown with IEEE sharing.

International level	**ISO** iso.org		IEC iec.ch		**IEEE*** ieee.org
National level worldwide	**DIN** din.de	**SAC** sac.gov.cn	**SA** standards.org.au	**ANSI** ansi.org	**JIS** jisc.go.jp (and further org.)
Europe	**CEN** cen.eu		**CENELEC** cenelec.eu		
National level in Europe	**ASI** austrian-standards.at	**UNI** uni.com	**DS** ds.dk	**SFS** sfs.fi	**BSI** bsigroup.com (and further org.)
	DIN din.de		**DKE VDE** dke.de		

* of special interest: IEEE (Institute of Electrical and Electronics Engineers), as one of the largest professional associations worldwide, has participated on the international level.

Figure 1: Standardisation organisations, international and national

The following overview in Figure 2 shows the time line of the work by JWG 16. By the way: Many of the participants in JWG 16 are working in an honorary capacity, which has had some influence on the composition of the committee.

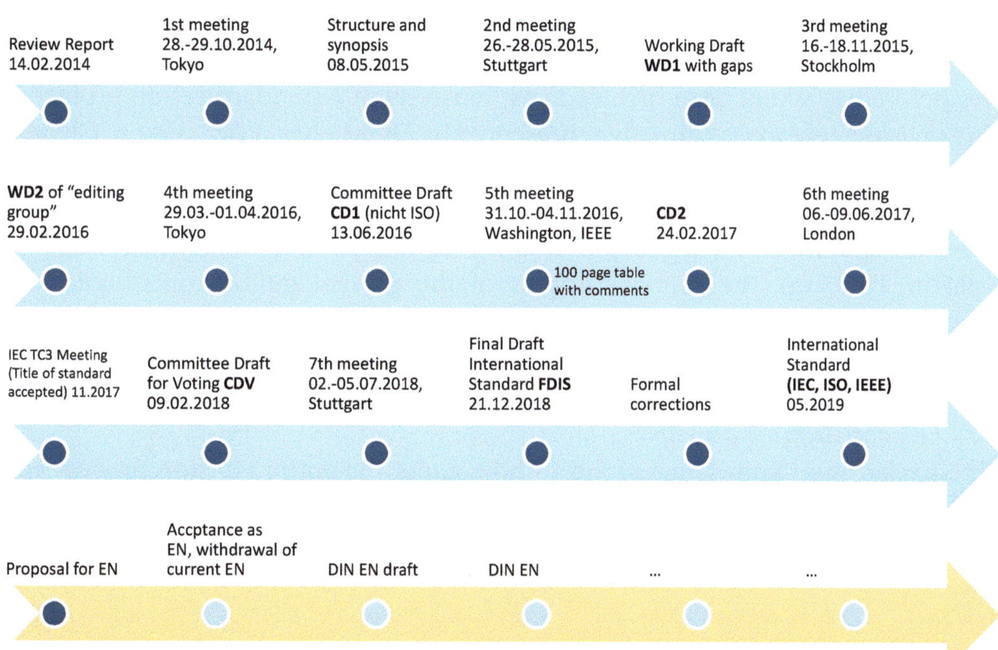

Figure 2: Time line of the standard´s revision

In all, 21 experts from nine countries participated in the revision effort. Additionally, experts from numerous further countries commented the draft standards and thereby contributed to the international applicability of the standard.

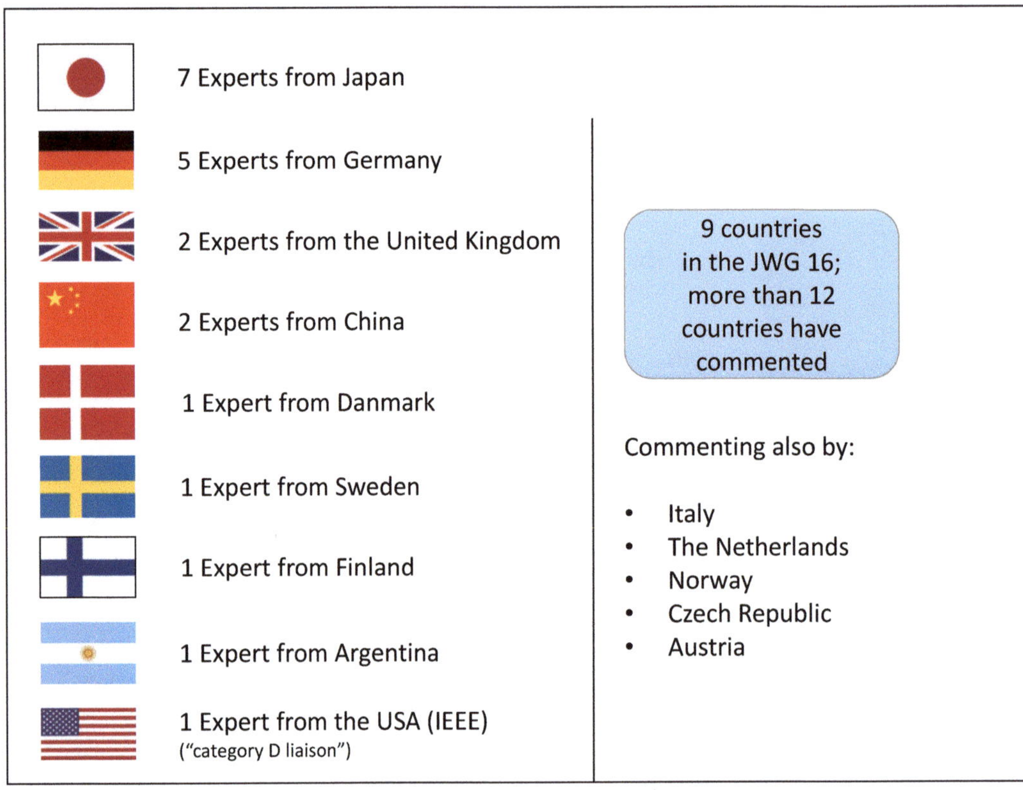

Figure 3: Composition of the Joint Working Group 16 by countries

1.3 Participation of tekom and the share of the authors of the present publication

Never before has tekom been involved as intensively in a standardisation project as in this revision. tekom was directly represented by Dr. Michael Fritz and Dr. Claudia Klumpp. In 2017, Dr. Klumpp assumed the convenor-ship, that is she acted as the chairperson of the working group. Roland Schmeling and Stephan Schneider shared in the international working group as leading experts.

Martin Tillmann, the primary author of the present guide, contributed significantly to the work of the German mirror committee for IEC/IEEE 82079-1, DEK GUK 113.1. Martin Rieder is an expert of the international working group ISO TC 199 WG5, that worked on *ISO 20607 Safety of machinery – Instruction handbook – General drafting principles.*

The background knowledge of the authors about the entire revision process and the numerous discussions, for instance regarding the development of the new concept of "information for use", has found its way into this practical implementation guide.

1.4 Who are in the target audience of this practical implementation guide?

The target audience is identical with the target audience of the standard itself: The practical implementation guide is addressed to all those who are responsible for or who share in the design, preparation, sustainment, translation, localisation, integration of content, production and delivery of information for use.

1.5 Normative contents and innovations

Upon restructuring, IEC/IEEE 82079-1 now contains the following sections:
(1) Scope
(2) Normative references
(3) Terms and definitions
(4) Fulfilment of requirements (previously: evaluation of conformity)
(5) Principles
(6) Information management process
(7) Content of information for use
(8) Structure of information for use
(9) Media and format of information for use
(10) Professional competencies
Annex A (informative): Guidance on evaluation (of fulfilment of requirements)

The scope outlined in *section 1* has not been extended, but defined in more detail. The normative references and definitions of terms in *sections 2 and 3* have been revised but—apart from the new concept of information for use—do not hold any great surprises.

Section 4 replaces section 7 (Evaluation of conformity) of the previous version. Based on ISO and IEC rules the section was renamed and moved up front.

Section 4, in discussing the options for providing evidence of fulfilment of the requirements of the standard, contains a pertinent novelty: A supplier cannot not only claim to have fulfilled the requirements for the information for use, but may also claim fulfilment of the requirements for the information management process. It is at the supplier´s discretion whether he wishes to claim to have fulfilled just the requirements for the information for use or also the requirements for the information management process. This serves to emphasise the importance of the new *section 6*, which outlines the requirements for the information management process. However, for consumer products fulfilment of the requirements for the information for use alone is relevant.

Scattered across the standard, already the first edition contained quality principles concerning information for use. These are now neatly collected and presented in clearer and more consistent wording in *Section 6*, under the heading "Principles". The "Minimalism" principle has been added, now resulting in at total of seven principles governing high-quality information for use:
- completeness
- minimalism
- correctness

- conciseness
- consistency
- comprehensibility
- accessibility

Section 6 is dedicated entirely to the information management process. The information management process is characterised by the principle of employing reproducible procedures.

The following four process groups are differentiated and dealt with in a subsection each:
- analysis and planning of information (subsection 6.2)
- design and development, including review, editing and testing (subsection 6.3)
- production and distribution (subsection 6.4)
- sustainment, maintenance and improvement (subsection 6.5)

A large subsection has been dedicated to analysis of the target audience(s). Analysis of the target audience is not a new requirement; however, so far only few organisations have a sufficiently meaningful target audience analysis, which can be used in developing appropriate information products. The detailed presentation of target audience analysis is meant to contribute to improving this situation.

The contents of information for use are covered in *section 7*. At first glance, many content items well known from the previous version are still listed. Viewed in detail, however, a large number of new items have also been added. For instance, the requirements for information about spare parts and consumables have been substantiated, and the requirements for the contents of a safety clause or section have been revised.

Section 8 is dedicated to the structure of information for use and is much more detailed than subsection 5.15 of the previous edition. For instance, the section recommends structuring information for use based on an information type model, and thus recommends following the three methods listed below:
- developing an information model by using a structuring method
- using an existing information model, for example an open source information model
- adapting an existing information model by using a structuring method

Section 8 provides for in-depth assistance in selecting suitable structuring principles. Additionally, this section describes the structure of step-by-step instructions concretely and comprehensively.

Section 8.4 is dedicated to navigating in and delivery of information for use. "Navigating" printed information for use continues to be covered. However, the focus now is on the aspects of dynamic delivery of information for use, e. g. pertaining to context sensitivity and search functions.

Section 9 addresses diverse means of communication, media, and the presentation of information for use; it contains specifications for durability, availability and presentation, including design of safety notes and warning messages.

Additionally, the external conditions of use have been addressed, to support the selection of the correct format or medium. For instance, if the target audience is service technicians for heating systems, who are often working in badly illuminated cellars, paper as a medium is not the optimum choice.

In summary, *section 9* contains numerous requirements already covered by the previous edition of the standard. One of those is the overview of recommended text font sizes.

Section 10, as a novelty, presents task and proficiency-related competencies for the designers/authors of information for use as well as requirements and recommendations for the competencies of translators. The first edition of the standard contained a short subsection (4.2) on this subject; however, the concept of absolute requirements for competencies has been dropped and replaced by process-oriented competency definitions. These are meant to support enterprises in developing the concrete requirements for the personnel involved in the information management process.

1.6 Structure of this implementation guide

Note:
The standard is subdivided into "sections" and "subsections". This implementation guide is subdivided into chapters. Therefore, references to "sections" refer to the standard, while references to "chapters" refer to the present implementation guide.

The implementation guide focuses on practical implementation of the standard. Since the structure of IEC/IEEE 82079-1 has been improved substantially, the structure of this implementation guide largely follows the sections of the standard. However, the implementation guide first addresses sections relating to the requirements for information for use itself; aside from some of the principles presented in *section 5*, these are mainly found in the *sections 7 to 9*. Thereafter the implementation guide deals with *sections 6 to 10* of the standard, which cover the information management process.

CHAPTER 11 THE RELATION BETWEEN IEC/IEEE 82079-1 AND PRODUCT OR SECTOR-SPECIFIC STANDARDS deals with the question of how IEC/IEEE 82079-1 can be implemented in parallel with product-specific standards and their requirements concerning information for use. One such pair of standards is dealt with in depth, the parallel implementation of IEC/IEEE 820791 and ISO 20607.

All chapters discussing specific sections of the standard begin with a table of "mandatory requirements" of the respective section of the standard. The subsequent sub-chapters then discuss the requirements and their implementation in detail.

1.6.1 Mandatory requirements and strong recommendations

The mandatory requirements form the major part of the requirements set forth in the standard. The focus of the present implementation guide on these mandatory requirements is meant to reach its objective, which is to help readers get going with implementation of the standard. That does not mean, however, that the strong recommendations of the standard are irrelevant; rather, those recommendations are quasi-requirements, and an organisation should have definite reasons to deviate from any of them, if it wishes to claim conformity with the requirements of the standard as a whole.

Standards use certain formulations to differentiate the types of requirements. The basis for this are the guidelines of the standardisation organisations. This implementation guide follows these guidelines. Table 1 provides an overview of the formulations used in the standard and this implementation guide.

Table 1: Formulations adapted to ISO/IEC Directives Part 2: 2018, section 7 and DIN 820-2, section 7

Requirements	
shall	is to is required to it is required that has to only ... is permitted it is necessary needs to
shall not	it is not allowed [permitted, acceptable, permissible] is required to be not is required that ... be not is not to be need not do not
Recommendations	
should	it is recommended that ought to
should not	it is not recommended that ought not to
Permissions	
may	is permitted is allowed is permissible
need not	it is not required that no ... is required

1.6.2 Use of the implementation guide

The practical implementation guide is meant as an introduction to the standard, helping to comprehend the normative contents. Since the structure follows that of the standard, it can easily be read together with the standard, helping readers understand its individual sections.

Thanks to its structure being aligned with the standard, the implementation guide is also most suitable as a reference work.

2 Pertinent changes and references to other standards

2.1 Paradigm shift: From document to information

The scope of edition 2 of the standard has been extended substantially. Previously, an integral instruction handbook or manual—a document in the sense of the standard—was the object of standardisation; but now the standard refers to "information for use". Such information is not necessarily presented in an integral document, but may come as part of the user interface of a software product, a catalogue, messages in an app, or a portion of a website on the internet.

This paradigm shift was needed primarily because modern electronic media, the presentation of information on monitors and screens as part of a growing number of products, as well as user instructions distributed by means of social media platforms have far outgrown the conventional "document" concept.

2.2 Title of the standard

The mind shift from document to information is such a central idea of the standard that it has made its way even into its changed title: The previous title read "Preparation of instructions for use – Structuring, content and presentation", whereas the new title highlights information: "Preparation of Information for Use (Instructions for Use) of Products". Since the title change amounts to a noticeable intervention into the standardisation landscape and must be well-founded, it was intensely discussed by the working group and the standardisation committee.

In parentheses the title still contains the term "Instructions for Use" to facilitate locating the standard in databases; the success of any standard depends on the acceptance it finds in the market, thus, visibility is an important criterion.

The title of part 1 of the standard (currently there is only this one, but further parts are planned) has likewise changed. The current title was "General principles and detailed requirements". In the new title of part 1 the phrase "detailed requirements" was dropped, instead it reads: "Principles and general requirements". The background of this change is the fact that further parts of this standards series are now in the making that will contain more detailed requirements for information for use in special cases. These parts will all be based on part 1. Currently, standards dedicated to assembly instructions and system/plant documentation are planned.

2.3 Scope

The scope of the standard (*section 1* "Scope") has not been changed in edition 2, at least not in principle. However, the substantially more detailed formulation—referencing "information" instead of "instruction"—clearly shows that it had to be clarified. The listing of some examples in the scope clearly shows that the standard is now applicable to the following products:

- machinery and systems/plants

- medical products
- consumer products
- software
- technical services

A wide range of media is explicitly mentioned in the scope of the standard, for instance:
- service information, provided as a PDF file
- online help for software
- troubleshooting information on a machine's on-screen display
- functional descriptions of a medical device on a website of the manufacturer
- printed information for assembly in the packaging of a consumer product
- planning information for a sensor, downloadable from a website
- tutorials as web-based training
- materials for on-site product training
- labels on a product

Videos supporting the use of products, thus also come in the normative range of the standard. Although the current edition only contains requirements concerning the synchronism between animation sequences and spoken or written text (see CHAPTER 9.3), additional requirements may be set forth in a further part of the standard series. The requirements for the information management process, the quality principles, and for the representation of text in videos are quite concrete and may be applied to video productions.

2.4 Information types and information management

By shifting the focus from discussing documents to contemplating information required, the standardisation committee opened the way to introducing the basic structure of information management to the standard. This is especially evident in the following parts of the standard:
- The introduction contains a graph illustrating how managed and classified information is made available by means of information products. The classification of information emulates the information model of DITA (**D**arwin **I**nformation **T**ype **A**rchitecture). However, this is not meant to limit information modelling to the model used by DITA.
- The principles include the single-source principle and require that information be centrally managed in a single location. The underlying reason is that redundant information management always increases the risk of inconsistencies.
- *Section 6* of the standard presents specific requirements for information management, particularly addressing configuration management, which highlights the need to include meta data in quality assurance.
- *Section 8* summarises further concrete requirements for information management.
- *Section 10* explicitly mentions competencies in information management and connects these to the fields of comprehensibility and product safety.

2.5 IEC/IEEE 82079-1 as a horizontal standard

IEC/IEEE 82079-1 has the status of a horizontal standard. According to IEC´s definition, a horizontal standard is a "standard on fundamental principles, concepts, terminology or technical characteristics, relevant to a number of technical committees and of crucial importance to ensure the coherence of the corpus of standardization documents". Already the scope highlights the status of IEC/IEEE 82079-1. It expressly mentions the relationship between the standard and sector-specific or product-specific standards such as IEC 60335 for household electrical products, ISO 20607 for instruction handbooks for machinery, and ISO/IEC 26514 for systems and software manuals, which is currently undergoing revision, and which is to be better aligned with IEC/IEEE 82079-1 in the future.

This sample listing of standards in the scope is by no means exhaustive. For instance, there are numerous standards on hand-held electric power tools containing detailed requirements for operating instructions; or consider ISO 20417 dealing with information to be provided by the manufacturer of a medical device.

IEC/IEEE 82079-1 shall be applied parallel to the respective sector-specific or product-specific standards and is to provide a well-founded base from the perspective of technical communication. Notably the principles and the requirements for the information management process will likely not be found in other standards.

The following CHAPTER 2.6 deals in greater depth with the relationship between IEC/IEEE 82079-1 and other sector-specific or product-specific standards.

2.6 Normative references

Section 2 "Normative references" of the standard lists standards that are referenced in other sections of IEC/IEEE 82079-1 in connection with certain requirements. If those requirements are mandatory, the standards referenced in *section 2* must be considered alongside. However, this should depend on the requirement in question.

Additionally, more than 50 other standards are listed in IEC/IEEE 82079-1, the reference to these is informative only. Informative references are not connected directly to fulfilment of a specific requirement and are listed in the bibliography at the end of the standard—not in *section 2* "Normative references". It is recommended to check the bibliography, to identify additional relevant standards.

The fact that a standard is not listed does not mean that it is irrelevant. At the same time the requirement for standards research should be taken seriously (subsection 6.2.5 "Information sources"). For instance, the US American standards series ANSI Z535 for presentation of safety notes and warning messages, warning signs and pictograms is not expressly listed, although the requirements in IEC/IEEE 82079-1 concerning safety notes and warning message are closely aligned to ANSI Z535 and are compatible with it.

3 Groups of important terms

Section 3 "Terms and definitions" of the standard does not contain requirements. Nonetheless, it is of vital importance when it comes to understanding the standard, because the terms used throughout the standard are to be understood as defined in *section 3*. The definition of terms may differ from the general or intuitive understanding and even from definitions in other standards and sets of rules.

To provide better overview of the defined terms, the definitions are here collected in the following groups of terms and are further explained where needed:

- documents and information products (information for use etc.)
- target audiences and user groups
- products (product, plant, system, component, and equipment)
- information structures (topic and information model)
- process
- safety
- quality

3.1 Documents and information products

Definitions contained: document (3.10), documentation (3.11), graphical symbol (3.13), illustration (3.16), information for use (3.17), information product (3.19), label (3.22), marking (3.24).

While the term "information for use" refers to any information within the scope of the standard, the terms "information product", "document" and "documentation" or "collection of documents" ("documentation" is used as a superordinate concept in the standard) refer to published information that is already embedded in a medium. A document is a 'fixed and structured amount of information intended for human perception that can be managed and interchanged as a unit between users and systems'. A video presenting instruction, thus can be considered a document. "Information product" is the most general term in this context; thus an information product may be a part of a document, a document itself, or even an entire documentation.

Graphical symbols are differentiated from safety signs, forming a special group of symbols.

3.2 Target audiences and user groups

Definitions contained: target audience (3.42) and user (3.47), competent person (3.3) and skilled person (3.36), consumer (3.7), customer (3.9), and supplier (3.39).

The term "target audience" is a central concept of the standard and is used about 100 times throughout the standard. Numerous requirements are related to the target audience, underlining its importance. The term "user" is broader in meaning: While user refers to a 'person who interacts with the product', target audience only refers to the 'group of persons for whom information for use is intended by the supplier'. However, the terms "user" and "target audience" are not used in the standard with a clearly defined demarcation.

3.3 Products

Definitions contained: component (3.4), consumable (3.6), consumer product (3.8), equipment (3.12), plant (3.27), product (3.28), self-assembly product (3.35), supported product (3.40), system (3.41).

The term "product" is a superordinate concept for all other terms in this field and refers to all 'results of labour, or of a natural or artificial process', even if the results are merely intended or projected.

The "system" is composed of 'interrelated elements considered in a defined context as a whole and separated from their environment'. A "plant" is an 'assembly of different systems on a specific site' and is generally referred to as a "complex system" in the standard (see subsections 7.10.11 and 7.13). To be able to describe the complex hierarchies of a plant, the term "subsystem" is used in addition to "system" and "component", however, it does not appear separately among the definitions.

3.4 Information structures

Definitions contained: information model (3.18), step-by-step instructions (3.38), topic (3.43).

Figure 1 in the introduction of the standard will help understand the term "information type".

Within the context of the standard, a "topic" roughly defined is a 'coherent part of an information product with a heading that deals with a single subject'. Within the frame of an "information model" each topic can be assigned to certain "information types". The term "information type" is introduced in subsection 5.2.1 and is also used in subsection 8.2 of the standard. Step-by-step instructions can be considered a specific information type.

3.5 Process

Definitions contained: commissioning (3.2), maintenance (3.23), repair (3.31), step (3.37), use (3.45), use case (3.46).

The term "use case" is used in the standard to refer to 'systems and their interaction with a user' and appears in connection with risk management (subsection 6.2.7) and the information types (subsection 8.2). In other parts of the standard, the interaction between the product and the user is described by the terms "task" and "goal".

The other terms do not need any additional explanation, but should be understood within the respective context.

3.6 Safety

Definitions contained: consequence (3.5), harm (3.14), hazard (3.15), personal protective equipment (3.26), product safety label (3.29), reasonably foreseeable misuse (3.30), risk (3.32), safety note (3.33), safety sign (3.34), usability (3.44), warning message (3.48).

These terms do not need any additional explanation, but should be understood within the respective context.

3.7 Quality

Definitions contained: accessibility (3.1), information quality (3.20), minimalism (3.25).

These terms do not need any additional explanation, but should be understood within the respective context.

4 Fulfilment of the requirements of the standard

Concerning the claim to fulfilment of the requirements for information for use the standard differentiates products for commercial use (B-to-B products) and consumer products (B-to-C products).

- For B-to-B products the manufacturer (or supplier) may decide to fulfil the more stringent requirements for the information for use, or the requirements for the information management process, or both.
- Because the requirements for protection of end users are generally higher for consumer products, B-to-C products must only fulfil the requirements for information for use.

Evidence of fulfilment of the requirements of the standard

Subsection	Requirement to be checked	Notes
4.1 Assessment	– For all products, except for consumer products: ▹ The requirements for the information management process in subsection 5.4, section 6 and section 10 are fulfilled. AND/OR ▹ The requirements for the information for use in sections 5.2, 5.3, 7, 8 and 9 are fulfilled. – For consumer products: ▹ The requirements for the information for use in sections 5.2, 5.3, 7, 8 and 9 are fulfilled.	
4.2 Evaluation	The section does not contain requirements, only recommendations.	
4.3 Documentary evidence	– The documentary evidence of assessment must comprise the following: ▹ the information provided and all packaging ▹ the language version that has been assessed ▹ information on whether the product to be assessed was supplied fully assembled or was assembled as part of the assessment of the information for use ▹ name of the organisation or person(s) that carried out the evaluation	

For assessment of the fulfilment of the requirements set forth in the standard, all information relating to the product shall be considered. That also includes all information for marketing of the product in any form (also in electronic form or presented on a website). Assessment of the user manual/instructions alone generally will not be sufficient to provide evidence of the fulfilment of the requirements; where applicable, a quick-start guide, the information on the packaging and on the product itself, the information displayed on a screen that is part of the product, instruction videos as well as information on websites, in catalogues and brochures must be considered. This is needed simply because contradictions between these information sources could lead to misunderstandings and, consequentially, faulty operation.

For consumer products, particularly, the standard recommends performing empirical effectiveness testing, which should involve persons unfamiliar with the use of the product. Annex A of the standard contains additional information on this topic. In practice, usability testing has also proven helpful for B-to-B products and their user manuals/instructions, as this can help in improving the user-friendliness of both the product and the instructions. This recommendation was already contained in the first edition of IEC 82079-1.

Further considerations revolve around the sufficient competency and independence of the persons performing the assessments.

Figure 4 illustrates the requirements of the standard concerning assessment of fulfilment of the requirements of the standard, dependent on the target audience of the information for use.

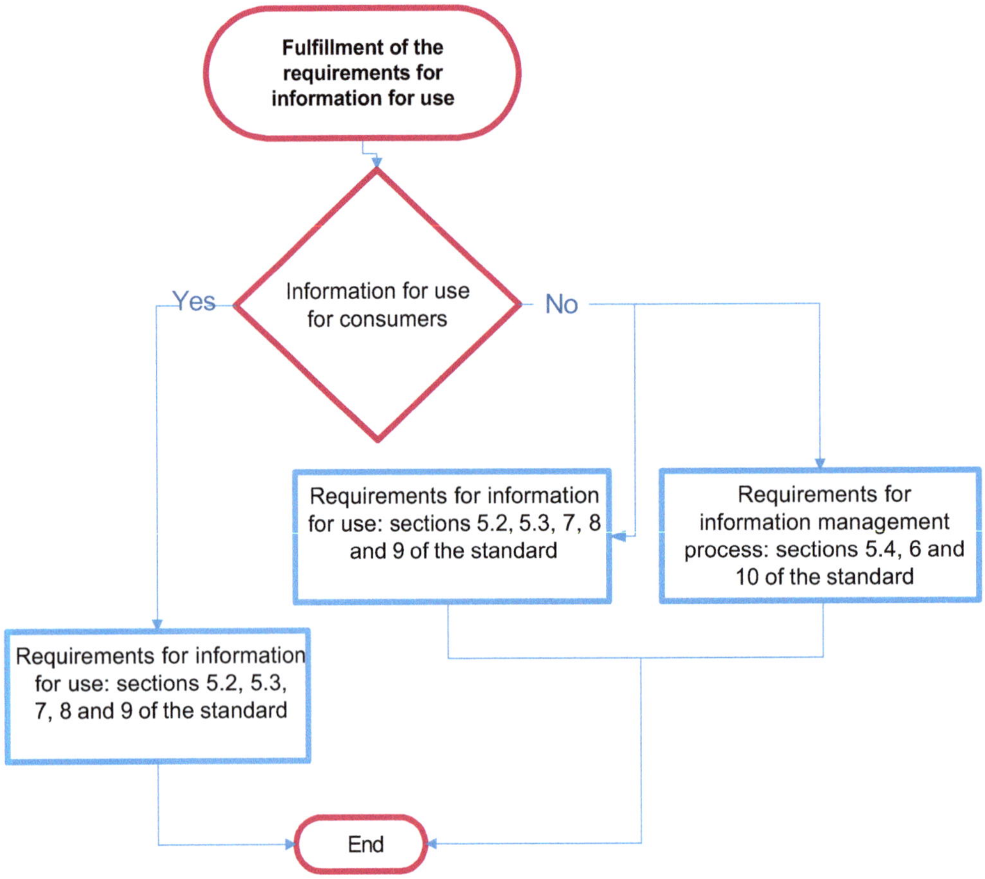

Figure 4: Providing evidence of fulfilment of the requirements of the standard

5 Principles

5.1 General

The principles play a central role in the standard, which is readily seen in the following quotation from annex A.2.2 of the standard: "Where fulfilment of a specific requirement is incomplete, but overall the information for use is judged as fulfilling Clauses 8 and 9, the reasons should be explained in the record, referring to the principles in Clause 5." That is, if detailed requirements of the standard are not fulfilled in their entirety, but the requirements as a whole can be considered fulfilled, fulfilment of the standard can be evidenced by referring to the principles. Thus, the principles in a way are considered superior to the more detailed requirements, and a deviation in some detail is acceptable, as long as the principles are adhered to.

List of requirements based on principles

Subsection	Requirement to be checked	Notes
Adherence to general principles		
5.2.1 General	– Information for use provides the necessary information for the target audiences to make safe, efficient, and effective use of the supported product	
5.2.2 Part of the product	– The information for use has been given the same attention and importance as every other part of the product – The information for use can be clearly and easily identifiable with the supported product.	
5.2.3 Target audience orientation	– The information for use is usable and relevant for the target audiences with respect to their expected tasks and goals	
5.2.4 Safe use	– The information for use promotes the safe operation and maintenance of the supported product	
5.2.5 Product compliance	This section does not contain any requirements, but only a statement	
Principles to be applied to ensure information quality		
5.3.2 Completeness	– The information for use is complete. To ensure this, the information for use must cover the following aspects: ▷ the risks of using the product ▷ the tasks that the target audiences are intended or allowed to perform throughout the life cycle of the supported product ▷ the target audiences' need for information ▷ all legal requirements, including ▷ the contractual requirements	
5.3.3 Minimalism	– Minimalism has been applied to the information for use – Safety-related information has been repeated where needed (as an exception from the minimalism principle)	
5.3.4 Correctness	– The information for use is technically correct – The information for use contains the current information on the supported product	

Subsection	Requirement to be checked	Notes
5.3.5 Conciseness	– The information for use is concise with respect to contents, format and media: ▹ wording is succinct (short and precise) ▹ illustrations do not contain unnecessary details ▹ videos contain only relevant material	
5.3.6 Consistency	– The information for use is consistent with respect to contents, format and media:	
5.3.7 Comprehensibility	– The information for use is comprehensible for the target audience(s): ▹ The text and terminology are comprehensible for the target audience(s) ▹ The illustrations, safety signs and graphical symbols are comprehensible for the target audience(s) ▹ The methods of navigation and use of media are comprehensible for the target audience(s)	
5.3.8 Accessibility	– The information for use is accessible for the target audience(s) in a concrete context of use – The information for use is accessible for the target audience(s) throughout the intended lifetime of the supported product	
5.4 Use of repeatable processes	This is a process-related requirement, see insertion 5.4 at the end of CHAPTER 6	

5.2 Purpose of information for use

5.2.1 General

To fulfil the requirements of the standard, the information for use must generally contain all necessary information enabling safe, efficient, and effective use of the product.

Safe: The information for use must contain all notes alerting to residual risks, which have been identified in course of risk assessment. For integration of safety notes and warning messages see CHAPTER 7.11.

Efficient: Efficiency is the ratio of the success of work to the effort invested. Information for use must be designed such that the user can successfully use the product with the least possible effort.

Effective: Effectiveness is the ratio of achievement of an objective to the definition of that objective. Thus, information for use must be designed such that the objective achieved matches the objective defined as closely as possible.

To meet these purposes, the information for use must provide the following information types:

- conceptual information: concepts, explanations and descriptions to enable the target audiences to perform tasks by understanding their purpose and the principles of operation of the supported product

- instructional information: procedures and task-oriented step-by-step instructions (regarding structuring also refer to subsection 8.3.4 of the standard and CHAPTER 8.3 of this implementation guide)
- reference information: detailed information on tasks that need to be performed occasionally only. This may include troubleshooting information

5.2.2 Information for use as part of the product

The standard requires that information for use 'shall be given the same attention and importance as every other part of the product'. Additionally, information for use 'shall be clearly and easily identifiable with the supported product' (see CHAPTER 7.2).

5.2.3 Target audiences orientation

The standard requires that information for use 'shall be usable and relevant for the target audiences with respect to their expected tasks and goals'.

To achieve that, target audience analysis is indispensable (see CHAPTER 6.2.2).

Note:
For additional information on target audience analysis refer to ISO/IEC 26514.

5.2.4 Safe use of the supported product

Safe use of the product is one of the most important principles for fulfilment of the standard.

Information for use must further safe operation/use (see CHAPTER 7.10.7) and safe maintenance (see CHAPTER 7.10.10) and contribute to reducing the risk of injury as well as the risk of inefficient use, malfunctioning, and damaging devices.

For the required contents of the standard refer to CHAPTER 7.

5.2.5 Product compliance through information for use

The standard states that information for use is part of the product. Within the framework of product compliance, information for use is generally relevant for the following areas of legal practice:
- product safety law
- product liability law
- contract law

5.3 Information quality

5.3.1 General

The standard requires that information for use meets the quality criteria completeness, minimalism, correctness, conciseness, consistency, comprehensibility, and accessibility. These requirements are discussed in more detail below.

5.3.2 Completeness

The standard says that to achieve completeness, information for use needs to address the following aspects:
- 'the risks of using the product'
  The respective information can primarily be retrieved from the manufacturer's risk assessment, performed parallel to the development, design, and manufacturing processes.
- all 'tasks that the target audiences are intended or allowed to perform throughout the life cycle of the supported product'
  The respective information can—aside from in the target audience analysis—be found in the risk assessment and/or product analysis of the manufacturer.
- 'the target audiences' need for information'
- The target audience analysis will contain information on this (see CHAPTER 6.2.2).
- all legal requirements, including contractual requirements

To ensure completeness, the entire organisation should be involved in the research. It is especially recommended to seek the participation of the research and development, design, marketing, and sales departments. It should be noted, that completeness may have safety-related aspects.

With a view to the contractual requirements, it should be checked whether adherence to the standard has been agreed upon, and if so, with which version. If the supplier switches to the newer or national version, the agreement with the customer should be updated accordingly.

5.3.3 Minimalism

According to the standard, information for use must be minimalistic. That is, the authors should refrain from long drawn-out explanations and filler words, as well as from redundancies. Only safety-related information may be repeated.

5.3.4 Correctness

According to the standard, information for use must contain current, technically correct information on the supported product.

To ensure fulfilment of this requirement, the product must be monitored and the information for use must be put under version control reflecting changes to the product.

5.3.5 Conciseness

The standard requires that information for use shall be concise with respect to content, format and media.

The following aspects are considered relevant concerning concise content and presentation:

- wording is succinct (short and precise)
  This requirement corresponds to the requirements for minimalism (see CHAPTER 5.3.3) and correctness (see CHAPTER 5.3.4).
- text and illustrations do not contain unnecessary details
- videos contain only relevant material

5.3.6 Consistency

As required by the standard, 'information for use shall be consistent with respect to the content, format and media'.

As interpreted by the standard, 'consistency with respect to content means, for example, that the information is unambiguous and correct' (see CHAPTER 5.3.4). Consistency with respect to format and media particularly means consistent use of terminology (see CHAPTER 7.5), units of measurement (see standards series ISO/IEC 80000), structures and indications (i. e. headings, titles), symbols, warning messages (see CHAPTER 7.11) as well as consistent use of colours (see CHAPTER 9.13).

As recommended by the standard, 'to achieve consistency, systems for consistent formatting should be separated from systems for consistent content'. Ideally, that means first gathering (consistent) content and then—separately—determining (consistent) formatting.

5.3.7 Comprehensibility

According to the standard, it is expected that the information for use is comprehensible for the target audience(s). To ensure that, the following need to be adapted to the needs of the target audience(s):

- the text and terminology used
- the illustrations, safety signs and graphical symbols
- the methods of navigation and use of media

The (culture-dependent) perception, user habits, and educational level of the target audience(s) must be considered to satisfy their needs.

5.3.8 Accessibility

The standard requires that the information for use 'be accessible (e.g. technically available and legible) for the target audiences in the context of use and throughout the intended lifetime of the supported product'.

To meet this requirement, it should be determined, as part of product observation, how the target audience will interact with the information for use supporting the product.

Note:
When providing information for use on websites, the supplier must develop strategies ensuring that the information for use remains accessible even if the website's address changes. This can be achieved, for instance, by programming redirection links.

5.4 Use of repeatable processes

Use of repeatable processes essentially refers to the information management process and is therefore covered by the reference to CHAPTER 5.4 at the end of chapter 6.

6 Information management process

List for inspection of fulfilment of the requirements for the information management process

Subsection	Requirement to be checked	Notes
Fulfilment of general requirements		
6.1 General	– An information management process for planning, designing, producing, and sustaining information for use has been implemented – The information management process generates, obtains, confirms, transforms, retains, retrieves, and disseminates information – The process ▷ identifies information to be managed, ▷ defines information representations, ▷ includes concepts to obtain, develop or transform, store, validate, present, and dispose of information, ▷ identifies the status of information, ▷ makes information available for use – The four process groups have been implemented: 1. Analysis and planning of information for use (section 6.2) 2. Design and development, including review, editing and testing (section 6.3) 3. Production and distribution (section 6.4) 4. Sustainment, including maintenance and improvement (section 6.5)	
Analysis and planning of information for use		
6.2.1 General	– Analysis and planning of information begins with an organizational strategy and policy regarding the development and sustainment of information for use – Instructional procedures are formulated based on a market analysis or an analysis of the characteristics, needs and intended tasks of the target audiences – The analyses are performed early in the life cycle of the supported product, and by collecting information about the product – Factors affecting the information design for information for use include target audience analysis, media and languages, information sources, risk management, and legal considerations	
6.2.2 Target audience(s)	– Planning of information for use considers the skills and tasks required of the target audiences – Planning of information for use considers the characteristics of the target audiences	
6.2.3 Media	– Analysis and planning of information for use include the choice of media for the information for use Example: Electronic and/or printed media, provided as part of the supported product, its packaging, or separately	

Subsection	Requirement to be checked	Notes
6.2.4 Languages	– Planning of information includes identification of local language requirements for the information for use – The local language requirements have been identified and versions of the information for use are translated and provided accordingly	
6.2.5 Information sources	– Planning of information includes the identification of information sources and the most effective ways to use them	
6.2.6 Information sustainment	– The needs for maintaining and supporting information during the life cycle of the supported product have been taken into account when planning the preparation of information for use – Information is managed so that relevant information on product changes and new releases is available for inclusion in the process – Planning for information for use includes identification of the expected duration and level of post-delivery support for the supported product and, therefore for the information for use. Plans to store, retrieve, change, protect, and dispose of the information for use have been implemented	
6.2.7 Risk management	– Analysis and planning of information is based on considerations of hazards inherent in the installation, operation, and maintenance of the supported product – The risk assessment (analysis of hazards) considers tasks or use cases – The supplier identifies, analyses, and evaluates hazards for their impact, likelihood of occurrence, and availability of methods to avoid or limit them – Residual risks (risk remaining after risk reduction) are included in the information for use as limitations or special exceptions and addressed in warning messages – The planning process includes the treatment of risks in the information for use	
6.2.8 Legal requirements	– Requirements regarding information for use resulting from contractual agreements and legal constraints are available for the information development process	
6.2.9 Project management and control	– Managing projects includes planning, developing, maintaining project schedules, performing resource management, and measuring and reporting the status of the project and the information for use – Project control includes the responsibility for authorizing and closing projects, obtaining and communicating accurate and timely information on project status, and redirecting project activities and tasks to take advantage of opportunities or to correct deficiencies	

Subsection	Requirement to be checked	Notes
6.2.10 Configuration management	– Persons in charge of configuration management ▷ establish baseline versions and manage changes to information for use under their control, ▷ control work information, such as reusable content ("single source") for information for use, product components, plans and design artefacts, ▷ define activities to review and approve changes and releases	
6.2.11 Human resource management	– Preparation of information for use includes planning for, obtaining, and allocating the required competent human resources – This includes training in the use of software tools and methods and familiarization with organizational procedures where necessary	
6.2.12 Quality assurance	– When preparing information for use, quality assurance techniques and activities have been identified and are implemented throughout the information development process	
Design and development, including review, editing, and testing		
6.3.1 General	– The development process includes the following: ▷ preparation of information design concepts, templates or models ▷ gathering of information ▷ selection of the content structure ▷ application of the information design to create the written and graphic content ▷ implementation of the information in the specified media; ▷ editing and review of the content, and evaluation of the information for use together with the supported product	
6.3.2 Design	The section does not contain requirements, only recommendations	
6.3.3 Review	– Errors discovered during reviews and testing are recorded, analysed for severity and time and cost to correct, and are resolved – A designated person is the review authority for determining whether the information for use has met its review criteria and can proceed to the next stage	
Production and distribution		
6.4	– Production of information for use comprises the integration, preparation, reproduction, packaging and distribution of physical media or electronic copies of the information for use	

Subsection	Requirement to be checked	Notes
Sustainment, maintenance and improvement		
6.5	– Sustainment includes continuing target audience feedback, in order to keep the information for use current for different versions of a supported product throughout the life cycle of those products. Additionally, feedback is used to make modifications with a view to improving usability – The supplier has established a method for receiving information on changes, updating information for use, and making updates available to the target audiences	
Process-related requirements from section 5		
5.4 Insertion concerning use of repeatable processes	– The processes ▹ are defined and repeatable ▹ are planned so as to obtain sufficient competent human resources for information development and to make information accessible at the required time ▹ guarantee that the quality is managed and assured on the basis of clear and common quality objectives ▹ use single-source information management	

6.1 General

The standard requires that the supplier develops and implements an information management process to create the information for use.

This topic had been handled in an informative annex in edition 1 and now is the subject of a substantial valorization due to it being treated in a normative section of edition 2. The requirement is primarily targeted at product manufacturers. However, suppliers may be involved in this information management process, particularly in delivery, monitoring and disposal of information for use. Information for use is part of the product and is thus subject to the obligation to monitor products in the market; this also must be fulfilled by suppliers.

The information management process:

- obtains information
  It determines how to perform product and target audience research, for instance.
- develops and stores information
  It determines how to archive the information obtained, for instance.
- validates information
  It determines how to validate the information obtained, for instance.
- transforms and distributes information
  It determines which media are used to distribute the information, for instance.
- retrieves information
  It determines which information can be retrieved by whom and when, for instance.

The process must ensure the following:

- Identification of the information to be managed
  It must be determined which information is to be passed on to those involved in creating the information for use in order for them to be able to create information conforming to standards. This includes:
  - ▹ risk assessments
  - ▹ target audience analyses
  - ▹ marketing documents
  - ▹ documents on similar products (own and those of competitors)
  - ▹ documents from service (e. g. concerning claims and customer queries)
  - ▹ documents from market research concerning own and competing products
- Definition of the presentation of the information
  It must be determined which information is to be presented in what form, e. g.:
  - ▹ quick-reference printed on paper
  - ▹ operating instructions in a file on a chip
  - ▹ troubleshooting as "help" on a display
  - ▹ information for maintenance and safety-related troubleshooting as an AR file (augmented reality file) of a qualified partner
- Concepts for sustainment, development, storage, change, validation, presentation and disposal of information as well as for determining the status and for distribution
  Concepts determining how to define the information development process must be developed and held available. Time lines, resources and budgets must be determined.
  It must be recorded, for instance, how information can be made available if the expected service life of a product exceeds the service life of the storage media. The concept should consider that the information must still be available even if the storage medium has been destroyed. The concepts for sustainment and storage of information are directly intertwined.
  It must be considered that the information for use may change, e.g. based on information from service, and the supported product may also be developed further. It is important to ensure that, on request, the users will receive the most recent version of the information for use matching exactly the product supplied to them. This requires the ability to query the status of the individual information for use published.
  The concept also determines how information for use is presented to the users. The presentation of information for use to a large extent depends on the target audience (see above).
  The creation of information for use is also subject to the needs imposed by sustainability. Therefore, the presentation of information for use must ensure that it can be disposed of in an environmentally compatible way at the end of use.

In designing the concept, the following four process groups shall be addressed:

- analysis and planning of information for use (see CHAPTER 6.2)
- design and development, including review, editing and testing (see CHAPTER 6.3)
- production and distribution (see CHAPTER 6.4)
- sustainment, including maintenance and improvement (see CHAPTER 6.5)

6.2 Analysis and planning of information

6.2.1 General

Analysis and planning of information for use must be part of a strategy aimed at establishing guidelines (specifications and processes) and organizational entities (structures) based on them, for the purpose of development and sustainment of information for use. In this, senior management should arrange the required resources. Senior management will also need to determine schedules and quality targets. This is needed to enable planning, preparation and delivery of information for use. To achieve this, senior management will need to meet their obligation to delegate responsibility and devolve authority to the relevant entities, typically to specialist departments.

Instructional procedures, particularly, shall be developed based on market and target audience analyses. The needs and characteristics of the target audiences and their intended tasks must be considered. Such analyses shall be performed early in the life cycle of the supported product by collecting information about the markets, the target audiences, and the product itself. The context of use, in which the product is used, likewise needs to be determined. Ideally this would include the operational concept of the product. The results of the analyses and their evaluation should lead to decisions concerning the following aspects:

- Will the information be attached directly to the product or supplied separately?
- Which media and means of communication will have to be used?
- What about the media concept? Will a monolithic information product be supplied or a series of documents containing the information for use? Will the information for use be supplied in paper format or digitally, too? If the information for use is to be supplied in digital format: will it be supplied on a storage medium or placed on the Internet, too? If a storage medium is used, considerations concerning future safety will bring up the question, which storage medium will ensure that the information remains available during all phases of the product life cycle.
- Will information have to be split up for and adapted to the needs of different target audiences?
- Which language versions will be needed?

In summary, it may be said that the information management process must lead to development of an information concept considering the following factors:

- target audience analysis
- media and means of communication
- language management
- access to information sources
- risk management
- legal requirements

6.2.2 Target audience

As stated earlier, planning and preparation of information for use requires considering the skills, needs, and the respective tasks of the target audiences. It should be noted that several target audiences with different tasks and expectations—active in different phases of the life cycle—will typically exist. These include the following groups of persons, for instance:

- users
- operators
- technicians
- installers
- start-up engineers
- administrators
- safety/security officers
- persons responsible for recycling and disposal

The supplier's assumptions concerning the skills and knowledge of the target audiences and regarding the frequency of use of the product should be determined and described.

As a result of the target audience analysis, the relevant characteristics of the different target audiences should be described. Common features can then be defined for the target audiences, for instance, concerning similar tasks. These must be considered when planning the information for use. The following is a list of relevant features to be addressed in the target audience(s) definition:

- education, skills, experiences, and training with respect to the tasks to be performed with the supported product (e.g. electrician, mechanic, specialist for hydraulics or pneumatics etc.)
- language of the target audience
- tasks of the target audience
- working environment of the target audience and the conditions of use of the information for use
- available tools and auxiliaries
- means for access to the information for use (e.g. Internet access, adequate devices for display of information as well as access rights)

The description of target audiences should consider that target audiences are often heterogeneous, which requires addressing a wide variety of skills and needs.

The description of the tasks of a specific target audience may consider the following aspects, for instance:

- Motivation for performing the task
  It should be considered, for instance, whether the user's objective is to ensure continuous operation of the product, to remedy a fault, or to perform (regular) maintenance. He may also wish to use the product for a short time only, for instance to cook coffee or to shave.
- Frequency of use
  Special consideration should be given to the question if it is likely that the user will remember how to perform the task. The less frequent a user will perform a

task, the more important it will be that the information needed to perform the task is readily accessible (as intuitively as possible).

- Time needed for the task
  The shorter a task, the less likely it will be that the user wishes to look for the information needed to perform it. Especially, for short safety-related tasks fast (intuitive) access to the information is important.
- Task performed by an individual or a team
  Providing information on tasks to be performed by a team certainly presents a challenge. This requires clear-cut communication concerning who—having what type of qualification—is to do what and when.
  A possible solution would be, for instance, to assign symbols to the different qualifications (see Table 2):

Table 2: Example of symbols assigned to qualification levels

	Administrator
	Operator
	Electrician
	Mechanic
	Maintenance staff

- Ambient conditions
  The supplier will have to research whether special conditions prevail while the task is performed and thus need to be considered in the information for use. For instance, a task may have to be performed under unfavourable ergonomic conditions, such as above one's head, or underground at dim light, or outdoors, where rain or snowfall and wind may occur. Additionally, stress may have to be considered, such as occurs in a piece-rate work environment or at a hospital.
- Fault tolerance
  For some tasks, correct execution is of vital importance; this particularly applies to safety-related tasks. These need to be addressed with special care at the information development stage. In most cases step-by-step instructions will proof beneficial; these should include intermediate result descriptions, providing the user with feedback on successful execution of the steps.
- Freedom of choice
  If a task could be performed by following alternative paths leading to the same goal, the preferred path should be described. But if a task can be performed

successfully by following one distinct path only, a step-by-step instruction will proof best.
- Prerequisites
  The prerequisites for the task must be described. These may include: use of appropriate tools, condition of the product (e.g. its surface temperature), standstill of interfacing products upstream or downstream etc.
- Commonness of the task
  Tasks that do not match the habits or experiences of the user will need special consideration at the information development stage. For instance, a user may have to be instructed not to extinguish a possible fire with water, because oil would continue burning and the contact with water would cause even greater damage.

Target audiences with a particular need for accessible or low-text information include among others:
- older persons
- functional illiterates
- children, who need to be supervised when using the product
- persons unable to read the information for use
- persons with physical, visual or auditory disabilities

Authors developing information for use for such target audiences will need special cognitive competencies.

6.2.3 Media

Analysis and planning of information shall include the choice of media and means of communication for the information for use. Table 3 provides an overview:

Table 3: Means of communication with reference to media

		Media					
		Attached to product	On packaging	Paper	Data storage medium	Internet	Presence
Means of communication	Nameplate	■					
	Bar-code or RFID	■	■				
	Text, in Braille also	■	■	■	■	■	■
	Symbols	■	■	■	■	■	
	Illustrations	■	■	■	■	■	
	Online help				■	■	
	Video, utilities movie				■	■	■
	Animation, possibly interactive				■	■	■
	Webinar					■	
	Training, CBT[1] and WBT[2]				■	■	■

[1] Computer Based Training
[2] Web Based Training

The supplier should select the media and means of communication in view of the needs of the target audience. Accessibility, applicability, and sustainability must be ensured. Information for use should be available and usable for the intended lifetime of the supported product. For example, printed information for use should be provided on durable paper. Electronic information for use can be provided in combinations of audio, visual (i.e. illustration, video, or animation), or in written format, on separate media, through the Internet, for use on various stationary or mobile viewing devices.

Note:
For design of a media concept refer to the tekom guideline "Delivery of information for use in electronic form – eDoc".

This overview goes to show that text is still considered the most important means of communication and may be used with all media. To simplify accessing information for users speaking a language different from that of the country of use and for functional illiterates, low-text or text-free means of communication may have to be used, if it is expected that this group of persons belongs to the target audience. This does not only apply to consumer products. Production equipment and other capital goods may include tasks performed by target audiences that may include persons speaking a language different from that of the country of use and functional illiterates, e. g. for transport, assembly or cleaning.

6.2.4 Languages

Generally, information must be supplied in the language(s) of the country of use (or sale). Additionally, planning of information shall include identification of the legal and actual local language requirements for the information for use. These may differ from the official language of the country of use (sale). In individual cases, contract agreements may require supply of additional language versions, e. g. for service staff. The translation process will then have to prepare and supply the required language versions.

Consideration should be given to conveying important installation and operational precautions and safety-related information by means of language-independent safety signs, graphical symbols and self-explanatory illustrations.

Note:
Refer to ISO 17100 for information on translation management processes.

6.2.5 Information sources

Planning of information shall include the identification of information sources and the most effective ways to use them. This requires intensive research. The following points should be considered:

- Function, structure and life cycle of the supported product (information source: e.g. research and development department)
  Similar products and requirements for information for use (information source: e.g. research and development department, technical editing).
  Possible problems and troubleshooting information (information source: e.g. research and development department, service and complaints management).
  Spare parts and options (information source: e.g. research and development department and service department).
- Market(s) in which the supported product will be supplied, including characteristics of the market(s) and channels for distribution of the supported product
  Information on this can generally be acquired from the sales and marketing departments.
- Information concerning compliance and conformance with legal and normative requirements (information source: e.g. research and development department, legal department and technical editing)
  If the product is meant for sale in the European market, the websites of the European Commission dealing with diverse product categories can be used for research. These sites do not just offer the text of the applicable directives and regulations (in all European languages), but also include guidance documents and a list of harmonised standards for download. If requirements for markets outside Europe need to be researched, it is advisable to engage a service provider in the research, who locates the applicable legal sources and summarizes the requirements found. International standards are best researched on the online browsing platform of ISO https://www.iso.org/obp/ui/#search.
- The time when stable and reliable information will be available

Below see a listing of which type of information about a product will reliably be available at what time:

Very early at the development stage
Information on contract agreements with reference to the information for use and information from the risk assessment (which will be continually updated/ improved throughout the project, however): intended use, technical data, limits of the product, target audience(s).

At the design stage
Detailed information from development and design concerning functionality. To gain such information, the technical communicator likewise needs to study the information on residual risks contained in the risk assessment. Additionally, information on materials used, consumables and substances required for operation must be gathered. The information for use dealing with operation, maintenance and disposal may be prepared based on this.

Shortly before market launch (prototype stage)
The developer of the information for use may study the product himself, receive 3D design files for processing and integration into the information for use; he may also receive information from the research and development department concerning first experiences with faults to be remedied.

Shortly before market launch, following usability testing
Experiences of a representative number of users with use of the product, considering the information for use developed at this point. This allows for the experiences made to be considered and the product to be launched together with the information for use.

Upon market launch
Information from product monitoring concerning actual use of the product (including faulty use) and from service. This information will not only be presented in the information for use, but will also influence further development of the product.

Reliable information concerning change of products or similar products already in the market should be researched at a very early stage. It should be considered how the new product differs from a predecessor or similar product. If the information for use for the similar product had been developed properly, design of the new information for use will be a relatively small effort.

The sources for information for use should also include querying persons and acquiring their observations concerning experiences made with the supported product. If information from subcontractors is included, the supplier of the product shall decide, if

- the subcontractor's information should be adapted by use of his own formats,
- he wishes to integrate the information into his own information for use by means of cross-references and hyper-links without making any changes to it
- he will add useful information about use of the subcontractor's product to his own information for use to precede cross-references.

6.2.6 Information sustainment

The standard requires that the needs for maintaining and supporting the information during the life cycle of the product is included in the planning for preparation of the information for use. For this purpose, information shall be managed so that relevant information on product changes and new releases is available for inclusion in the process. This is to include plans concerning how to store, retrieve, change, protect, and dispose of the information for use. Additionally, the expected time period for preparation and delivery of the information for use must be considered in the planning process.

This requirement includes a challenge. When the concept for sustainment of the information for use is prepared, some of the required information will not be available. Thus strategies for different scenarios will have to be drawn up and the concept will have to be adapted continuously along the product life cycle.

6.2.7 Risk management

Analysis and planning of information for use shall result in a concept for handling of safety-related information. The entire product life cycle shall be considered. Typically it includes transport and storage, assembly and installation, commissioning and operation, service and maintenance, troubleshooting and shutdown, recycling and disposal. The hazard analysis and risk assessment should include all tasks and use cases. Hazards include not only immediate risks to operators and users of the supported product, but also foreseeable future impact on others in the environment. Hazards also include the possibility of unfavourable or unexpected risks for the customer, such as loss of business or personal data. The supplier shall identify, analyse, and evaluate hazards for their impact, likelihood of occurrence, and availability of methods to avoid or limit them. Residual risks (risk remaining after risk reduction) shall be included in the information for use as limitations or special exceptions and addressed in warning messages.

The planning process shall include the treatment of risks in the information for use. This should be coordinated with the engineering and design departments to help ensure safe and proper use of the supported product. With respect to safety, the developer of the information for use should be involved in risk assessments so as to make clear the limits of information for use in ensuring safety. Over-warning should be avoided. Technical measures take priority over safety-related instructions. Generally, the following procedure must be ensured:

- Risks should be eliminated by design measures.
- Risks that cannot be eliminated or reduced sufficiently by design, shall be reduced by suitable safeguards.
- Suitable warning message shall warn of residual risks still remaining.

The process of iterative risk reduction is illustrated in fig. 1 of EN ISO 12100.

Only if neither inherently safe design nor safeguards can eliminate or reduce the risk, shall suitable safety notes and warning messages warn of the residual risks.

This imposes the following obligations on the developer of the information for use:

- Evaluate the residual risks together with the research and design department and determine the measures to be taken concerning safety notes and warning messages. In some cases, one may critically scrutinize whether a safeguard or inherently safe design measure may be chosen to eliminate the risk
- Select and formulate warning messages
- Classify the warning messages to determine the risk level and allocate the signal word
- Supplement the safety clause if needed

6.2.8 Contractual agreements and legal constraints

Requirements regarding information for use resulting from contractual agreements and legal constraints shall be available for the information development process. To facilitate this, the developer of the information for use must be informed of contractual peculiarities as early as possible.

6.2.9 Project management and control

Managing projects shall include planning, developing, maintaining project schedules, performing resource management, and measuring and reporting the status of the project and the information for use.

This should include the following aspects, if possible:

- preparation of and adherence to guidelines for content management and reusability
- review and testing of information for use
- translation and localisation for all target markets of the supported product
- management of sufficient resources
- involvement of competent persons (see CHAPTER 10)
- quality assurance, e. g. by usability testing (see Annex A of the standard) and configuration management (see CHAPTER 6.2.10)
- provision of infrastructure resources, such as a content management system with interfaces to the project management
- for complex systems (e. g. a processing plant), a delivery plan for the different supplier documentations

The standard requires that project control includes the responsibility for authorizing and closing projects. To achieve that, communication must be organised early and information on the project status be provided. Only this way, can weak spots be identified and remedied early.

6.2.10 Configuration management

The standard requires that the persons in charge of configuration management establish baseline versions and manage changes to information for use under their control. Reusable content (single source) for information is to be monitored. Additionally, the persons in charge must keep updated any information concerning product components, plans and concepts as well as have in place measures for review and approval of changes and releases.

The standard recommends that the persons in charge manage traceability of the versions of the information for use. It should be possible to make the current or applicable version of the information for use available to the target audience(s) at any given time.

6.2.11 Human resource management

Preparation of information for use shall include planning for, obtaining, and allocating the required competent human resources. This shall include training in the use of correct methods, familiarization with organizational procedures and use of the appropriate software tools where necessary. The required qualification of the developers is outlined in CHAPTER 10.

6.2.12 Quality assurance

When preparing information for use, quality assurance techniques and activities shall be identified and performed throughout the information development process. Notes concerning quality monitoring are found in CHAPTER 6.6.3.

6.3 Design and development, including review, editing, and testing

6.3.1 General

The standard requires that research of information, design of concepts, templates and models for the information for use shall be part of the design and development process. It also requires selecting an appropriate structure for the information for use.

Section 8 of the standard (see CHAPTER 8.4) deals in detail with structuring of information for use, particularly outlining suitable methods and criteria. The information design must be prepared and then applied. The contents of the information for use must be edited and reviewed. Additionally, the standard explicitly requires that the information for use be evaluated together with the supported product. The implementation of the information for use in the media previously selected likewise is part of the design and development process.

The standard recommends performing usability tests (see Annex A of the standard). Such tests may demonstrate that the information for use allows the target audience(s) to perform the tasks with the supported product safely and efficiently. During such testing, weak points may surface, helping to improve the quality of the information for use.

6.3.2 Design, information gathering and development

The standard recommends following a consistent design concept for preparation of information for use. To facilitate this, a style guide or guideline for authoring and editing should be established and followed throughout the preparation process by all persons involved. The following contents are recommended for such a guideline:

- requirements for
  - ▹ writing style and wording
  - ▹ terminology
  - ▹ communication among the team
  - ▹ design and page layout, including typography rules
    It should be noted that different requirements may have to be set for the different publication media.
  - ▹ review regarding fulfilment of the requirements
- rules on how to properly address the target audience
  For instance, one might determine that commands for actions to be performed with the product shall be written in the imperative ("Press the button to start the machine"), but not in passive voice ("The button is pressed to start the machine")
- rules for navigation and linking in the information for use (for minimum rules see CHAPTER 8.4)
- rules for information types, e. g. types of information for normal operation, maintenance and disposal

Note:
See ISO 704 for further information on the principles and methods of terminology work.

If translation is required, information showing how terms have been translated consistently for the target language shall be provided. If translations are prepared by an external service provider, the terminology defined internally shall be provided when the translation is returned to the editing department.

6.3.3 Reviewing, editing and testing

The standard recommends that reviewing and testing of the information for use should be performed throughout the information development process. To achieve this, methods should be established that ensure evaluation by independent persons. Special issues, such as terminology and adherence to an information model, may be best evaluated by experts. The entire review should be performed by an interdisciplinary team. Upon review, an evaluation of the information for use should be carried out by a panel best representing the target audiences (also refer to Annex A of the standard).

The source text should be reviewed for accuracy, consistency, and usability before it is translated. In addition to proofreading by a copy editor, use of a language checking software is recommended.

Empirical evaluation (e.g. usability testing) should be used as a means for determining whether the information for use achieves its purpose. Table 4 lists examples of empirical methods:

Table 4: Empirical evaluation methods

Empirical method	Scope and objectives of the method
Interviews for opinion survey	– Analysis of satisfaction by those involved in the preparation or presentation of the information for use – Target audience analysis – Identification of information regarding the context of use of the product (for example why, where, how often) – Identification of target audiences' needs – Assessment or evaluation of information for use – Comparison of different information products
Focus groups for opinion survey	– Identification of target audiences' needs – Assessment or evaluation regarding handling of and working with the information for use – Comparison of different styles and formats
Usability testing	– Identification of gaps and weaknesses in the information for use – Testing of information for use or comparison of different versions – Testing of comprehensibility and usability
Self-assessments and check lists	– Assessment of quality according to technical criteria – Check lists relating to standards
Expertise, expert reviews and certification	– Assessment of information for use according to defined criteria – Analysis of weak points – Attention to legal concerns
Feedback, recommendations, or observations from independent award or merit schemes	– Evaluation of a product according to generally accepted guidelines – Comparison with other information products
Feedback from complaints management, hotlines and service information	– Continuing improvement process – Optimization of request handling – Improvement of the information product

Table taken from IEC/IEEE 82079-1, Table 1: Examples of empirical methods

The standard recommends incorporating improvements achieved by the evaluation into the style guide in order to continually improve the quality of information for use. To meet the requirements of the standard, errors discovered during reviews and testing must be recorded, analysed for severity and cost to correct. A person must be designated who is responsible for review and for determining whether the information for use has met its review criteria. It goes without saying that any deficiencies found must be corrected.

The standard also recommends considering the time for reviews and correction of significant defects in the project plans. If necessary, the need for repeated reviews should be considered.

6.4 Production and distribution

The standard requires that production of information for use comprises the integration, preparation, reproduction, and distribution of the different physical media. Aspects of logistics should be considered. The better a manufacturer knows his target audience and is capable of supplying directly to the target audience, the easier it will be to meet this requirement. When distributing information for use by means of content delivery systems, the correct chunks of information can be supplied using suitable classifications (taxonomies), that is by input of meta data arranging information in categories or classes (taxa).

6.5 Sustainment, maintenance and improvement

The obligation to monitor products also applies to the information for use. This must include continued collecting of target audience feedback, in order to keep the information for use current for different versions of a supported product throughout the life cycle of the product. It would also include feedback on modifications relating to improved usability.

The standard requires that the supplier establishes a method allowing to make current updates available to the target audiences. This requirement proves to be challenging especially where information for use is supplied in the form of printed documents.

It is comparatively easy to keep electronic information for use updated, e. g. by means of content delivery systems. In this connection the standard recommends considering information on customer satisfaction, customer complaints, service reports, and information on accidents when preparing amendments of the information for use. This should include informing the target audiences via the available media of safety-critical revisions to the information for use made after sale of the product.

When it comes to optimising the information for use, the standard explicitly allows considering information published by third parties, e.g. in You-tube tutorials or blogs. This possibility is purposely mentioned by the standard as neither a requirement nor a recommendation, because the amount of information about a product provided by third parties may be unmanageably large (this may especially apply to consumer products).

Insertion 5.4 Use of repeatable processes

The use of repeatable processes required in subsection 5.4 of the standard is a process-related requirement and is therefore discussed here in CHAPTER 6.

According to the standard, information for use must be developed, produced and sustained using defined, repeatable processes.

These processes must

- be planned so as to obtain sufficient competent human resources for information development and to make information accessible at the required time to the persons involved,
- be subject to quality management and quality assurance and
- use single-source information management to reduce inconsistencies and discrepancies

The standard recommends retaining the information content for reuse throughout the intended lifetime of the supported product, as well as to keep it updated. The information content should be held in a form making it adaptable for use with similar products and for future product versions.

7 Content of information for use

Note:
The information does not need to be developed or presented in the order here given.

List of requirements for content of information for use

Subsection	Requirement to be checked	Notes
Fulfilment of general requirements		
7.1 General	– The information for use contains the information required to satisfy the needs of the target audience(s) concerning safe, effective and efficient use of the product (identical with CHAPTER 5.2.1) – The principle of minimalism has been applied (identical with CHAPTER 5.3.3) – The information for use describes the intended use of the product – The information for use helps the target audience(s) ▹ avoid unacceptable safety risks, ▹ malfunction, ▹ damage to the supported product itself or ▹ inefficient operation – The information for use helps identify and avoid reasonably foreseeable misuse	
Identifiers		
7.2.1 Information for use	– The information product is unambiguously identifiable – The information product is identified by a unique identifier and reference information	
7.2.2 Product	– The information for use clearly identifies the supported product – The information for use contains the unique identifier of the supported product (**where applicable**) – **If** the information for use addresses more than one product version: ▹ the identifier of the product version is clearly recognizable on the product and in the information for use	
7.2.3 Supplier	The information for use indicates the supplier of the product: – name and address – contact data ▹ **If offered:** contact data ▹ of providers of special tools ▹ of providers of materials ▹ of authorized service centres for the product ▹ of technical assistance	

Subsection	Requirement to be checked	Notes
Retaining of printed information for use		
7.3	– **If the** information for use is delivered in paper form: ▷ the information for use is to draw the attention of the target audience to the importance of retaining it during the lifetime of the supported product – the supplier makes available replacement copies of information for use (for example, from a website) throughout the intended lifetime of the supported product	
Presentational conventions		
7.4	– The information for use lists and explains the presentational conventions used	
Terminology, explanation of acronyms, abbreviations, technical terms, safety signs, graphical symbols, and markings		
7.5 Terminology	– Consistent terminology has been used throughout the information for use, on the product and on its packaging – The terminology is unambiguous – **If the** target audience will not hold specialist knowledge: ▷ the information for use explains the terminology used – **If the** information for use of a system (e.g. a processing plant) contains documents from various other suppliers using different terminology: ▷ the information for use explains the meaning of the differing terms so that the users can easily relate to the different terms in the different documents	
7.6 Acronyms, technical terms, abbreviations	– The information for use lists and explains all ▷ unavoidable technical terms, ▷ acronyms and ▷ abbreviations, **as far as these** cannot be readily understood by the target audience	
7.7 Explanation of graphical symbols	– All safety signs, graphical symbols, pictograms, markings, and icons on the product itself or the packaging have been explained	
Product description		
7.8.1 General	– The target audience can understand the essential functions of the product in the context in which it is intended to be used – The intended use has been described clearly and unambiguously – The technical limits and conditions within which the product is intended to be used have been considered	
7.8.2 Visualization	The section does not contain requirements, only recommendations	

Subsection	Requirement to be checked	Notes
7.8.3 Specifications	– The product description contains the following specifications, **where applicable:** ▹ complete measurements, e. g. of mass, volumetric capacity, and performance ▹ information regarding electrical ratings (e. g. input voltage); ▹ information regarding energy consumption, energy efficiency rating, insulation category, and IP code (as specified in IEC 60529) ▹ communication interfaces and protocols ▹ gas supply specifications (e. g. type, pressure), water supply and other relevant information on items such as cleaning agents, lubricants, and fuses (e. g. type, rated value and characteristics) ▹ emitted noise level, waste discharge rate etc. under specified conditions ▹ electromagnetic compatibility ▹ description of the type of intended user and, especially in the case of accessible design products, a clear statement of any special precautions that should be followed by the user ▹ standards, and legal requirements with which the supported product complies ▹ notes on intended use and reasonably foreseeable misuse of the supported product ▹ description of the user controls and indicators applicable to the relevant component(s); ▹ acceptable environmental conditions, such as ambient temperature or humidity ▹ information about the year of manufacture, expiration date, or limitations on hours of use, if supported products have an intended limited lifetime	
Information on supplied accessories, consumables and spare parts		
7.9.1 Accessories supplied	**If** accessory parts are required: the information for use contains the following information for each accessory part supplied: – name, part number – illustrations to allow the target audiences to identify the accessory and where to attach it – description of function and purpose – information needed to install the accessory – information needed to order a replacement	

Subsection	Requirement to be checked	Notes
7.9.2 Consumables	**Where** consumables are required: the information for use contains: – information concerning the recommended quantity – description of the consumables – the supplier's model name or part number and a generic name – illustration showing the position of, or access to, the consumables (information on how to remove, replace or refill the consumables) – information about disposal of consumables – information about when a consumable is no longer usable	
7.9.3 Spare parts	**If** the product features replaceable parts: – a spare parts list is provided	
Information for use needed during the lifetime of the supported product		
7.10.1 General	The information for use: – contains clear information about the respective restrictions, where legal requirements, safety, or technical reasons prohibit the installation, operation, or maintenance of the supported product by non-skilled persons – contains clear information concerning required qualifications and competencies	
7.10.2 Re-packaging	**Where applicable:** – the information for use includes information on special precautions to be taken for re-packaging	
7.10.3 Transportation and storage	**Where applicable:** – the information for use includes information for protection of the product and the safety of persons during transportation and storage	

Subsection	Requirement to be checked	Notes
7.10.4 Installation	**Where applicable:** – the information for use answers the following questions concerning installation: ▹ On the outside of the packaging: How is the product to be unpacked and the packaging to be disposed of safely? ▹ What is to be done, if the product itself or the packaging is damaged? ▹ What is to be done, if the packaging has been opened unintentionally before use? ▹ How shall transport and packaging restraints and fastening clamps be removed? ▹ Check list: What is included in the package? ▹ What is the minimum space needed for use, maintenance and repair? ▹ Layout plan ▹ Interconnection diagram ▹ Which legal restrictions, procedures and precautions apply to connection to energy and water supplies, drainage and other auxiliary supplies? ▹ Conditions for assembly and installation, including fastening (earthquake-resistant if applicable) ▹ How to check whether the product is properly installed and ready to perform as intended? – The information for use will contain appropriate information about the needed expertise, if the product is supplied to locations that prohibit the installation by non-skilled persons	
7.10.5 Commissioning	– The information for use includes the information necessary for commissioning of the supported product prior to its handing over in working order.	
7.10.6 Modification	**Where** modification is not admissible: – the information for use contains a clear statement concerning this restriction, together with information about possible consequences resulting from modification **Where** modification is permitted: – the information for use clearly describes and illustrates how to undertake the admissible modifications **or** – directs the target audience(s) to a source of relevant information	
Operation of the supported product		
7.10.7.1 General	The information for use provides information: – for normal operation – for exceptional situations – for emergency situations	

Subsection	Requirement to be checked	Notes
7.10.7.2 Normal operation	The information for use includes: – illustrations that clarify or reinforce understanding of the main functions and safety precautions – information on when and how the supported product should be used – information about local or remote operation and how to switch between the two – information for automatically controlled and remote-controlled products – information about manual operation and automatic operation and how to switch between the modes of operation – information about means of starting and stopping the supported product's operation – information on initiating operations, checks, adjustments, or conditioning required prior to intended use – information for identification and correction of malfunctions – recommendations for waste disposal Information regarding operational controls explain the following type of status indicators: – normal operation – indications that the system has automatically switched to safe operation mode before a hazardous situation has been reached – the current state of operation – what to do when the control system has failed – indication of whether the system is in manual, local, or automatic mode	
7.10.7.3 Exceptional and emergency situations	The information for use shall cover the following: – sequence of actions to be taken in the case of an unexpected interruption of power, cooling, control functions, or another emergency situation – fault indication and its location – restarting after an intervention – type of fire fighting equipment to be used in the case of fire – foreseeable emissions, leaks, or decomposition of harmful substances, together with associated warnings of the likely hazards and precautions for avoiding harm – first aid treatments recommended for foreseeable acute medical conditions that can result from the use of the supported product, and advice on subsequent professional medical treatment to be sought	
7.10.8 Indications of faults and warning device signals of the supported product	– Indications of normal and abnormal operation and the recognition of signals from any warning devices have been explained and can be identified. Such information is accessible without the supported product being in operation – Signals or displays intended to indicate impending risk of damage or imminent danger are readily comprehensible and unambiguous	

Subsection	Requirement to be checked	Notes
7.10.9 Meanings of signals	– Meanings of signals, such as flashing indicators, colour changes or audible signals have been explained so that they are easily recognized and understood by the target audience	
Maintenance of the supported product by non-skilled and skilled persons		
7.10.10.1 General	– Information for maintenance that can be performed by non-skilled persons is presented separately from information for skilled persons	
7.10.10.2 Non-skilled persons	**Where** non-skilled persons are allowed to perform maintenance: – the information for use describes those tasks with appropriate illustrations and an explanation of the nature and frequency of those tasks – regular maintenance tasks (e. g. daily), periodic (short-term and long-term) maintenance tasks and inspection activities are differentiated – materials and tools used for cleaning, for example, appropriate chemical substances, cleaning cloths and brushes, are specified – possible risks arising from the use of inappropriate materials and tools are clearly specified – maintenance tasks that should not be attempted by non-skilled persons are clearly identified. Information on how to contact the supplier or others to obtain technical assistance is included	
7.10.10.3 Skilled persons	– Information on maintenance by skilled persons includes: ▹ nature and frequency of inspections ▹ safety precautions and warnings for maintenance carried out on running or live equipment ▹ drawings and diagrams enabling maintenance to be done effectively ▹ information on regular checking of warning devices ▹ details of cleaning methods ▹ information on the required tools and materials ▹ schedules for planned periodic maintenance, or preventive maintenance based upon the results of the condition monitoring of the systems and components involved, or any combination of both methods ▹ contact details of the supplier or others from whom technical assistance can be obtained	
Maintenance of complex systems		
7.10.11 Complex systems	– Maintenance information includes the following: ▹ safety precautions for performing maintenance (such as complete or partial shutdown) ▹ procedures for accessing subsystems or components to perform maintenance, including means of emergency exit	
Troubleshooting and repair of the supported product by non-skilled and skilled persons		
7.10.12.1 General	– Information for tasks that can be performed by non-skilled persons is presented separately from information for skilled persons (separate information products or topics)	

Subsection	Requirement to be checked	Notes
7.10.12.2 Information for non-skilled persons	**Only where** non-skilled persons are permitted to resolve faults or repair of the product without incurring risk of harm to themselves or of damaging the product: Information has been provided for – methods of repair and, – **if needed,** methods for testing after repair	
7.10.12.3 Information for skilled persons	**Only where** skilled persons are permitted to resolve faults or repair the product: The following information has been provided – methods of repair and, – **if needed,** methods for testing after repair **Where applicable** – List ▹ of indications for fault identification and its location ▹ indications of normal operation – descriptions of built-in diagnostic systems to aid fault detection – drawings and diagrams to enable troubleshooting to be carried out effectively – Notes ▹ on how to detect malfunctions of the equipment ▹ on abnormal behaviour of the equipment ▹ on abnormal symptoms and indications ▹ on protocolling of alarm and trip conditions ▹ for start of standby or alternative systems ▹ for shutting down and isolating units, which do not operate as intended – On repairs and adjustments – Aids ▹ for troubleshooting procedures ▹ for step-by-step fault diagnosis of algorithms ▹ for fault trees and computer-based fault diagnosis in complex systems – Clear definition of troubleshooting and repair tasks that should not be attempted – Contact details of the supplier or others from whom technical assistance can be obtained	
Replacement of parts of the supported product by non-skilled and skilled persons		
7.10.13.1 General	– Information for tasks that can be performed by non-skilled persons is presented separately from information for skilled persons (separate information products or topics) – Contact details of the supplier or others from whom technical assistance can be obtained have been provided	

Subsection	Requirement to be checked	Notes
7.10.13.2	**Where** replacement of parts by persons is permitted without incurring risk of harm to themselves or of damaging the product: The following information has been provided: – method of replacement – **if required,** method for testing after the replacement – source of spare parts – replacing or placing of labels, **as far as** this may be needed during the expected service life of the product	
Disassembly, recycling, disposal of the supported product		
7.10.14.1 General	– The information for use includes information on disassembly, recycling or disposal of the product	
7.10.14.2 Disassembly	– **If** disassembly by the target audience is permitted, information on this has been provided, taking into account safety and environmental considerations	
7.10.14.3 Recycling	– **Where** specific procedures are necessary for recycling of the product or its components, this information shall be provided in accordance with the appropriate legal requirements and standards	
7.10.14.4 Disposal	– When providing information on disposal, safety and environmental protection are appropriately taken into account **Only for** products containing data or intellectual property: The information for use identifies: – responsibilities concerning disposal – determination of ownership of stored data and IT assets – procedures to remove and retain or to destroy data and IT assets – procedures to confirm that data has been removed from data storage devices before they are recycled, or to physically destroy the devices	

Subsection	Requirement to be checked	Notes
Safety-related information		
7.11.1 Types of safety-related information	– Three types of safety-related information are differentiated: ▹ safety signs and product safety labels ▹ safety notes ▹ warning messages – Safety-related restrictions concerning the use of the supported product are given – Safety-related information is incorporated into the information for the intended safe use of the product and includes the following: ▹ intended environment ▹ type of application ▹ tools needed ▹ protective features of the product to be installed or activated ▹ safe use and disposal of the product ▹ working materials and consumables ▹ identification of permitted modifications of the product ▹ procedures to maintain the product in a safe condition and to dispose of it safely	
7.11.2 Location	– Safety notes have been provided in a separate clause or section at the beginning of the information for use – The clause/section containing the safety notes: ▹ is clearly identifiable ▹ has a heading which emphasizes the importance of the content – In step-by-step instructions, warning messages are given in the context in which a hazard is likely to occur – **If** personal protective equipment is necessary: ▹ information for the personal protective equipment is provided at the beginning of the task description that requires the equipment ▹ information about personal protective equipment is given on the product package and the product itself, **where applicable**	
7.11.3 Target audiences	– Hazards for particularly vulnerable target audiences are provided at the beginning of the information for use – Information about hazards for particularly vulnerable target audiences are highlighted for emphasis	

Subsection	Requirement to be checked	Notes
Safety notes		
7.11.4 Safety notes	– Safety notes are indicating the following, as a minimum: ▹ possible hazards ▹ their causes, **if appropriate** ▹ information on how to avoid the hazards ▹ the probable consequences of not avoiding the hazards – The safety notes start with a general statement that highlights the importance of reading them – The safety notes include the following information: ▹ identification of hazards with appropriate graphics, including: · reasonably foreseeable misuse · inadvisable modifications of the product · when the product is no longer safe to use due to effects of wear or damage · radiation, including any sealed hazard, for example, sources of ionizing radiation, laser (in accordance with IEC 60825-1), microwave, ultraviolet, infrared and lethal voltages · substances which are carcinogenic, mutagenic or reprotoxic or other substances for which there is scientific evidence of probable hazard to human health · emitted sound pressure levels ▹ harmful potential consequences of unsafe use ▹ safety precautions (means of avoiding hazards), such as protective clothing or goggles; restrictions on the personnel who can use the product; observance of product controls and messages; safety signs and labels; safekeeping of keys and passwords; ▹ means of recovery from hazards, including emergency procedures and use of emergency equipment	
Warning messages		
7.11.5.1 General	– Warning messages contain the following, as a minimum: ▹ potential hazards ▹ their causes, **if appropriate** ▹ how to avoid the hazards ▹ probable consequences of not avoiding the hazards – Each warning message is placed either before the first step or incorporated just before the relevant step of a step-by-step instruction – Warning messages on how to avoid hazards are concise	

Subsection	Requirement to be checked	Notes
7.11.5.2 Design	– Design of the warning messages is consistent – Warning messages are made conspicuous and prominent – The following has been taken into account: ▹ warning messages begin with a signal word ▹ text and illustrations are limited to the essentials ▹ the location, content and style of the warning messages are conspicuous ▹ warning messages are visible to the target audiences and to any other persons exposed to hazards, from their position during use, and at the right time ▹ warning messages have been placed as determined by the risk assessment	
Signal words		
7.11.5.3 Harm to persons	The section does not contain requirements, only recommendations	
7.11.5.4 Damage to property	The section does not contain requirements, only recommendations	
Safety-related information in quick-start guides		
7.11.6 Quick-start guides	– Quick-start guides include safety-critical information relevant to the tasks described in the quick-start guide – Quick start guides make clear to the target audiences why and when they need to read additional safety-related information in the information for use – Quick-start guides include a reference to the safety-related information in the instructions: ▹ the reference is unambiguous and presented conspicuously at the beginning of the quick-start guide together with the general warning sign or a signal word ▹ the text of the reference describes the possible consequences of not complying with the general safety messages	

Subsection	Requirement to be checked	Notes
7.12 Self-assembly products	– The instructions for assembly of self-assembly products are presented in the sequence of assembly actions in the form of one-step-at-a-time sequence of illustrations, each accompanied by text positioned adjacent to the respective step – Where necessary, explanatory or advisory information and warnings are provided – The information for use specifies instructions for each assembly step as follows: ▹ the components and other objects involved in the step ▹ how components are to be aligned and oriented ▹ the actions the assembler needs to perform to shape, attach or connect the components ▹ the appropriate tools required ▹ safety precautions to take or foreseeable errors to avoid during the step ▹ how to confirm correct mating of components and that their attachment has the required strength – **If** assembly or erection require special skills, this shall be clearly indicated in the information for use by a warning such as "for assembly by a qualified electrician" or "assembly by adults required" A further requirement relates to the process: – The risk management (see CHAPTER 6.2.7) deals with foreseeable errors of the assembler that probably compromise the safety of the person undertaking the assembly or the safety of the completed product	
7.13 Complex systems	– Each system as a whole includes information for use, including restrictions for use and safety notes on the overall system level – The step-by-step instructions concerning operational tasks performed in a system are structured in a way to present the users' tasks, and do not only describe the relationship of the system components	
Information security and data privacy		
7.14.1 General	**Where** systems contain data: – the information for use includes instructions for features to protect the confidentiality, integrity, and availability of data from loss or misuse For remotely controlled or network-connected products: – the information for use includes features to avoid misuse or unauthorized access (hacking), such as changing default passwords	
7.14.2 Functions for access control or protection of sensitive data	– The most common methods of providing information protection or assurance have been taken into account when creating, exchanging, delivering or storing information (encryption, backup, data masking, and data erasure)	

Subsection	Requirement to be checked	Notes
7.15 Training	**Where** training is required for the target audiences:	
	– the information for use points to the need for such training	

7.1 General

The standard requires that information for use covers the needs of the target audience(s) for the safe, effective and efficient use of the product. This requirement is identical with the requirements in CHAPTER 5.2. Further information is found in that section.

Additionally, the principle of minimalism must be applied. Information on this principle is found in CHAPTER 5.3.3.

According to the standard, the information for use must describe the intended use of the product. In particular, information for use must help the target audience avoid unacceptable safety risks, malfunction of or damage to the product, or inefficient operation. Information for use shall provide the necessary information to identify and avoid reasonably foreseeable misuse. Further information on this is found in CHAPTER 7.8 and CHAPTER 7.11.1.

7.2 Identifiers

7.2.1 Identification of information for use

The standard requires that the information is unambiguously identifiable. The following are mentioned as examples for such identification:
- title
- identification number
- date of issue
- revision number (version) and issue date of the revision
- identifier for the language
  (ISO 639-2 contains the standard codes for identification of languages)

This requirement corresponds to the requirement that the information product must uniquely identify the product (see CHAPTER 7.2.2).

7.2.2 Identification of the supported product

The standard requires that information for use clearly identifies the product. To achieve that, a serial number range, identifying the information product, lot numbers and information about the configuration and other peculiarities of the supported product may be useful. A direct connection between the information for use and the product may also be achieved by the use of illustrations. Such information may appear on the title page of a printed information product for instance, as well as in (living) column titles placed in the page header or footer. For information products supplied in electronic format such information may be useful in the form of meta data and/or given on a start screen.

Where more than one product version is addressed by the information for use, the different versions shall be clearly recognizable in the information for use and on the product itself.

7.2.3 Identification of the supplier

Name and address of the supplier are mandatory contents of the information for use of the supported product. The standard additionally requires contact details such as telephone numbers, email addresses, and hyper-links. Other means of communication such as a fax number, QR or barcode may help with the identification.

If tools and materials from third party suppliers are needed for use of the product, the contact data of these suppliers as well as of authorized service centres for technical assistance shall be included.

7.3 Importance of retaining printed information for use

The standard requires that information for use provided in paper form draws attention to the importance of retaining the information for use for the intended lifetime of the supported product. The standard contains two examples of how to this:

Example 1:

IMPORTANT
READ CAREFULLY BEFORE USE.
KEEP FOR FUTURE REFERENCE.

Example 2:

	For safe and proper use follow these instructions. Keep them for future reference.

Figure 5: Examples similar to those in IEC/IEEE 82079-1 (7.3, Figure 2)

These and similar texts may serve to get the attention of readers.

Example 1 is taken from EN 82079-1:2012, however, it does not meet the requirements of ANSI Z535.6 for legibility (because of the all capitals text). Therefore it is advisable to follow example 2 in IEC/IEEE 82079-1:2019.

The supplier is additionally required to provide replacement copies throughout the intended lifetime of the supported product, for instance for download from his website.

To allow users to request a specific replacement copy, the requirement in CHAPTER 7.2 for identification of the information for use and the supplier.

7.4 Presentational conventions

The standard requires that presentational conventions used in information for use be listed and explained where necessary.

That is (at least) the structure of warning messages, step-by-step instructions and cross-references must be described. However, if all presentational conventions are used consistently, the users will readily comprehend and recognize them without explanations. To follow the principle of minimalism one should consider, which presentational conventions need to be explained (see CHAPTER 5.3.3).

7.5 Terminology

The standard requires that consistent, unambiguous terminology be used throughout the information for use, on the supported product, and on its packaging.

Terms, which are likely unknown to the target audience, should be avoided (see CHAPTER 5.3.7) or be explained, for instance in a glossary.

In information products for complex systems (e.g. a plant), which are referring to information products issued by other suppliers, using different terminology, the meaning of the differing terms must be explained so that the target audience can easily relate the different terms.

In printed information products this may be done, for instance, by inserting a "list of equivalent terms" at the beginning of each documentation provided by a different supplier.

Table 5: Example of a terminology equivalence list

Main term	Term used by supplier
gas burner	burner
chimney	smoke stack
thermolysis	pyrolysis Remark: This term is not quite correct for the chemical process occurring in the system, since the objective is to produce clearly defined products (petrol, diesel, LPG, and paraffin).

In electronic information products differing terminologies may be displayed from meta data upon "mouse-over" actions.

7.6 Acronyms, abbreviations, technical terms

Where unavoidable technical terms, acronyms (abbreviations composed of the initial letters of the abbreviated words) or abbreviations are used, the standard requires that these are listed and explained, unless they are readily understood by the target audience.

The preparation of a glossary is recommended for this purpose.

7.7 Explanation of safety signs, graphical symbols and markings

The standard requires that all safety signs, graphical symbols, pictograms, and markings placed on the supported product and the packaging be explained.

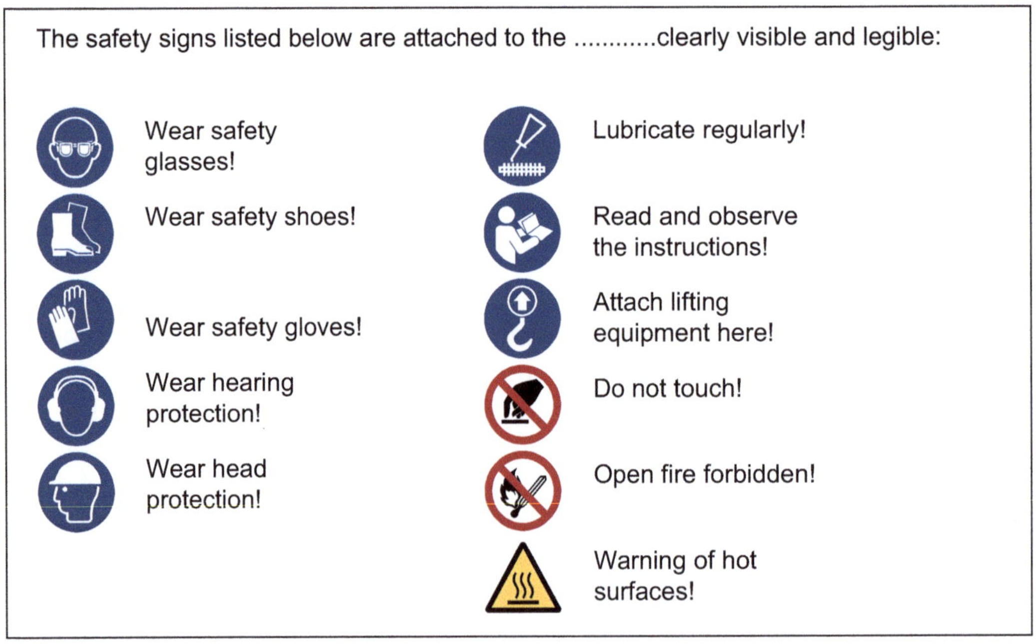

Figure 6: Example for explanation of pictograms

7.8 Product description

7.8.1 General description

According to the standard, the target audience needs to understand the essential functions of the product in the contexts in which it is intended to be used. Additionally, the intended uses of the product must be clearly and unambiguously described, the technical limits must be considered.

Ideally, the information on intended use and reasonably foreseeable misuse (required in CHAPTER 7.1) originate from the risk assessment.

To determine the intended use, it is best to first define the limits of the product. According to EN ISO 12100 these include limits of use, space limits, time limits, and other limits.

The operating modes are part of the limits of use. These may include normal operation, automatic operation, manual operation, or setup operation. The different phases of the life of the product, such as operation, maintenance, and troubleshooting must be considered. Additionally, the range of application, the target audience, and other persons that may be exposed to hazards from the product are determined.

Space limits mainly include the space required by the product itself and for interaction of operators with the product.

The time limits include the service life of the product and recommended maintenance intervals.

Other limits of products may be related to materials, cleaning, and environmental aspects.

All these details must be considered when determining the intended use of the product (also refer to CHAPTER 7.11.1).

Reasonably foreseeable misuse of products may be known from product monitoring or from experiences with similar products, which are already in the market. The description of reasonably foreseeable misuse entails implicit instructions of the target audience concerning what to do. This may to some extent contradict the requirement for comprehensibility (see CHAPTER 5.3.7). Generally, it is best to tell the target audience in a straight forward way how to act (formulated in active voice).

The requirement for a description of the reasonably foreseeable misuse originates, as one of several sources, with the Machinery Directive, annex I, section 1.7.4.1. However, a sentence like the following is not recommended, because it does not contain additional information, "Misuse includes all applications that are not described as intended use".

It is important to place a warning message before the description of a concrete action that might be carried out in a faulty way due to the person seeking the most comfortable way to do it, due to time pressure or false intuition (see CHAPTER 7.11.5). This is required in the Machinery Directive, annex l section 1.7.4.2.

Note:
For a list of all requirements in the Machinery Directive regarding information for use, "Instructions", refer to CHAPTER 12.1 *(Annex).*

7.8.2 Visualization

The standard recommends providing a visualization of the supported product. This is to help the target audience identify the product and its main functions.

For more details on this refer to CHAPTER 9.11.

7.8.3 Specifications

According to the standard, the product description is to include the specifications of the supported product. Where applicable, these include the following as per the standard:

- Technical data, such as complete dimensions, mass, electrical ratings and information regarding the consumption of electric energy or other energy sources (e.g. gas, oil, petrol, hydrogen)
  For some products minimum requirements from directives and/or type C standards must be observed. For instance, for products coming under the Machinery Directive (2006/42/EC) and the so called Outdoor Directive (2000/14/EC) the emitted sound pressure levels must be given.
- Information concerning communication interfaces with other products and protocols used to exchange information
- Information on electromagnetic compatibility, especially for products meant for use in sensitive areas (e.g. medical products)

- Information on the intended use and reasonably foreseeable misuse of the supported product (see CHAPTER 7.8.1)
  Environmental conditions (such as outdoor use or use in closed rooms, use in explosive atmosphere) may belong to the information on intended use. Whether it is better to list the intended use among the technical specifications, depends on the product and the concept of the information for use.
- Description of user controls and indicators
  This information should be included at the beginning of the information for use (following the safety notes), to allow users to familiarize themselves with them from the outset. If a touch screen is available, the information may be displayed context sensitive.
- Information concerning limitation of the operating hours, the service life and the expiration date
  This information may be placed in the information unit for the phase of life "disposal". The information unit for maintenance may also contain information on the limitation of operating hours until the next maintenance.

These content-related requirements of the standard regarding description of the product also apply to machinery coming under the Machinery Directive (2006/42/EC). To ensure completeness, all content-related requirements of the Machinery Directive are listed in the ANNEX.

7.9 Supplied accessories, consumables, spare parts

In summary, the standard requires that the information for use includes all information needed for ordering supplied accessories, consumables and spare parts.

Additionally, all information needed for replacement or installation, respectively, must be provided. The individual requirements for the minimum contents are listed below.

7.9.1 Accessories supplied with the product

Where accessories, consumables or spare parts belong to the product, the standard requires to include all of the relevant information in the information for use.

The standard requires integration of the following information on accessories in the information for use:

- Designation of the accessory and/or part number
- Illustrations to allow the target audiences to identify the accessory and order it
  In the case of paper documents a drawing with numbered captions placed counterclockwise around the drawing are recommended. If the information for use is supplied in electronic form, the accessories may be identified by pop-ups or may in some cases even be placed directly in a purchasing cart by clicking.
- Description of the function of the accessory
- Description of safe and efficient installation of the accessory. Where needed, information on testing of safe function

7.9.2 Consumables

The standard requires that the information for use contains the following information on the consumables of the supported product:

- Information on the recommended quantity, e. g. related to operating hours
- Sources (information on vendors) of consumables with precise product designation or other identifications
- Illustrations showing the replacement position of the consumable (preferably in an exploded view)
- Information on safe and efficient replacement or refilling of consumables
  Where required, notes from safety data sheets (e. g. concerning personal protective equipment) must be integrated.
- Information on legal disposal of materials
  In most cases, such information can be found in the safety data sheets. If that is the case, one should refer to the safety data sheet, since the rules for proper disposal of a consumable may change during the product life cycle.
- Information on the durability of the consumable

7.9.3 Spare parts

If a product has replaceable parts, a list of spare parts must be supplied as a minimum.

The spare parts list should contain all relevant ordering information.

7.10 Information for use needed during the lifetime of the supported product

7.10.1 General

Where the tasks to be performed during a phase of life of the product for technical or legal reasons require specially qualified personnel, the standard calls for a clear definition of such requirements in the information for use. The required qualifications and competencies for such tasks must be clearly stated.

Such information may be presented neatly in a "Who-does-what"-matrix (audience mapping matrix).

Table 6: Example of a who-does-what-matrix (audience mapping matrix)

Tasks/phase of life	**Groups of persons**			
	Transport carrier[1]	Assembly fitter[2]	Operator[2]	Maintenance technician[2]
Transporting	X			
Loading and unloading	X			
Set-up		X		

Tasks/phase of life	Groups of persons			
	Transport carrier[1]	Assembly fitter[2]	Operator[2]	Maintenance technician[2]
Connecting		X		
Start-up		X	X	
Operation			X	
Decommissioning			X	
Dismantling		X		
Troubleshooting			X[3]	X
Repair			X[3]	X
Service			X[3]	X
Maintenance			X[3]	X
Inspection			X[3]	X
Disassembly		X		
Disposal	X	X	X	X

[1] Specialized service provider
[2] Trained personnel
[3] Non-safety critical tasks (see manufacturer's risk assessment)

A prerequisite for drafting of such a matrix is a well-founded target audience analysis (see CHAPTER 5.2.3 and CHAPTER 6.2.2). For additional information on target audience analysis refer to ISO/IEC 26514.

7.10.2 Re-packaging of the supported product

To fulfil the standard, the information for use must include the special precautions to be taken when re-packaging, if a product must be re-packaged, e. g. for further shipment or storage.

7.10.3 Transportation and storage of the supported product

If the risk assessment shows that information for safety of persons during transport or storage is needed, such information must be integrated into the information for use.

7.10.4 Installation of the supported product

For products that are not supplied in a ready-to-use state and need to be unpacked, installed, or assembled, the standard requires that the information for use includes the following information:

Table 7: Information for installation

Information	**Location of the information** (not normative)
What to do if the product itself or the packaging is damaged	On the outside of the packaging and in the information for use, if needed
Notes and procedures for unpacking and disposal of the packing materials	On the outside of the packaging
What to do, if the packaging has been opened unintentionally before use	On the outside of the packaging
Notes on removal of transport and packaging restraints and transit clamps	On the outside of the packaging
Check list of items included in the package	Inside the package, in the information for use, if needed
Minimum space needed for use, maintenance and repair	On the outside of the packaging and in the information for use, if needed
Layout plan <DIESE ZEILE IST DOPPELT EINE LÖSCHEN>	Inside the package, in the information for use, if needed <DIESE ZEILE IST DOPPELT EINE LÖSCHEN>
Layout plan	Inside the package, in the information for use, if needed
Conditions for assembly and installation including fastening (e. g earthquake-resistance considerations, see CHAPTER 7.12)	Inside the package, in the information for use, if needed
Legal requirements, procedures and precautions for connection to the power and water supply, drainage and other auxiliary supply sources	Inside the package, in the information for use, if needed
Notes on check-up of proper installation and readiness for safe use.	In a check list, in the information for use, if needed

Additionally, the standard requires that the information contains an appropriate note, if the product is supplied to locations that prohibit the installation by non-skilled persons. The standard also requires that the supplier specifies the required specialist knowledge.

7.10.5 Commissioning of the supported product

The standard requires that the information for use contains the information necessary for commissioning of the supported product.

For complex machinery and plant systems a check list may be useful. Depending on the complexity, the check list may have to be subdivided for different phases of the life of the product.

Examples:

- initial operation
- daily (re)start
- re-start upon cleaning
- re-start upon remedy of fault
- commissioning upon prolonged shutdown or maintenance

7.10.6 Modification of the supported product

The standard requires that where modification of a supported product is not permitted, this restriction shall be clearly stated in the information for use. Depending on the consequences of such modification, a warning message may be needed (see CHAPTER 7.11).

Where modification of a supported product is permitted, it must be clearly described and illustrated how to undertake such modifications, to ensure it will be possible to continue using the product safely and efficiently upon modification. The information for use may point to an external source of information. This may be especially recommendable for extensive modifications or for modifications rarely made.

7.10.7 Operation of the supported product

7.10.7.1 General

Quite generally, the standard requires that the information product contain information on normal operation, on exceptional situations, and emergency situations.

7.10.7.2 Normal operation

For normal operation, the standard provides additional details concerning the above requirement. The following contents are required for normal operation:

- Illustrations that clarify the main functions and safety precautions
  These will include control cabinets, operating consoles, emergency stop devices, stationary and movable guards, signalling devices etc.
- Information about the intended use (see CHAPTER 7.8.1)
- Information about the operating modes, e. g. local operation, remote control, manual operation, automatic operation and setup operation
  The information should define which target audience is authorised to use which operating modes (e. g. by means of access authorization or keys), which operating mode takes priority (e. g. local operation over remote control) and what must be considered when switching between operating modes.
- Information about tasks for preparation and inspections to be carried out before commissioning (see CHAPTER 7.10.5)
- Information about correction of malfunction
  In most cases this will likely be the key point, as the target audience will have to be able to locate information on troubleshooting quickly (and as intuitively as possible), to be able to resume operation.
- Information about legally compliant disposal
  Information should not just include disposal of the supported product itself at the end of its service life but, in some cases, also about disposal of waste products generated by the product (e. g. metal shavings and waste cooling water of a turning machine).
  It may be necessary to refer readers to the regulations for disposal, if these might change during the product life cycle.

Additionally, information about indicators showing the operating mode of the product must be included, if the product has such indicators. Indicators for operating modes may include the following, for instance:
- operating mode (e. g. normal operation, shut-down, manual operation, automatic operation, setup operation, local operation, remote control)
- fault indication, emergency stop or start-up of operation (audible and/or visual)
- notes concerning needed action displayed at a panel (e. g. in case of failure of the control system)

Note:
For software documentation for user guidance systems, refer to IEC 61506.

7.10.7.3 Exceptional and emergency situations

If the risk assessment for the supported product reveals that exceptional situations or emergency situations may occur, the standard requires that the following information must be supplied (as a minimum):
- Sequence of actions to be taken in the case of an unexpected interruption of power, cooling, control functions, or other emergency situations
- Fault indication and fault location
- Restart upon intervention (also see CHAPTER 7.10.5)
- Type of fire fighting equipment to be used in the case of fire
  For large systems it will be useful to represent the escape routes and fire extinguisher locations in a schematic. Such information should not only be included in the information for use, but also readily visible on the product itself.
- Foreseeable emissions, leaks, or decomposition of harmful substances, together with associated warnings of the likely hazards and precautions for avoiding harm
  Information on this should also be given at the system itself, by means of pictograms as a minimum.
- First aid treatments recommended for foreseeable acute medical conditions that can result from the use of the supported product, for example, toxic shock, epileptic seizure or burns; advice on subsequent professional medical treatment to be sought

7.10.8 Indications of faults and warning device signals of the supported product

If the product features fault indicators and warning device signals, the standard requires to explain these in simple terms; the information must also be accessible when the product itself is not in operation.

One possibility would be to attach information sheets near warning devices, explaining the visual and/or audible signals. Such information should by all means be included in safety-related training regularly administered.

7.10.9 Meanings of signals

The standard requires that in addition to fulfilment of the requirement in CHAPTER 7.10.8 *all* signals, such as flashing indicators, colour changes, or acoustic signals must be explained in the information for use in a readily understandable form.

Note:
Examples of signals are found in IEC 60073, ISO 7731, ISO 11429, ISO 15006 and ISO/TR 16352.

7.10.10 Maintenance of the supported product by non-skilled and skilled persons

7.10.10.1 General

To satisfy the need for target oriented writing (see CHAPTER 5.2.3 and CHAPTER 6.2.2) the standard requires that information for maintenance by skilled and non-skilled persons is presented separately.

7.10.10.2 Maintenance by non-skilled persons

If non-skilled persons will be able to perform maintenance safely and efficiently, the information for use must include suitable descriptions of those tasks with appropriate illustrations. Detailed and illustrated step-by-step instructions with the intermediate results to be achieved and the end result will serve that purpose well. This will allow the non-skilled persons to check the results of their actions. This especially requires describing all prerequisites, such as the tools and materials required, and the condition of the product (e. g. device is switched off and secured against switching it on again). The information must alert to possible risks in an appropriate way (e. g. by warning messages, see CHAPTER 7.11).

Maintenance tasks that may only be performed by skilled persons must be specifically identified. Notes on how to contact skilled persons for such tasks shall be included.

Such maintenance tasks should not be included in the "normal" information for use, but be described separately.

7.10.10.3 Maintenance by skilled persons

According to the standard, information for use describing maintenance tasks to be handled by skilled persons must contain the following information (as a minimum):

- Nature and frequency of inspections
  To identify the frequency of inspections, it may be practical to include a chronological table of tasks to be done daily, weekly, monthly or yearly.
- Safety precautions and warnings (see CHAPTER 7.11)
  Such precautions are required only for tasks that must be performed while the product is operating or live. However, it is recommended to additionally consider notes concerning the maintenance phase contained in the risk assess-

ment. For instance, the product may have to cool down before maintenance, or components in the maintenance area may have to be set to a safe condition.
- Inspection of warning and safety devices
- Needed tools and materials as well as prerequisites concerning the operating condition
- Contact details of the supplier or others from whom technical assistance can be obtained

For further notes refer to subsection 7.10.10.3 of the standard.

7.10.11 Maintenance of complex systems

The standard requires that maintenance information for complex systems, such as industrial plants, includes the following information:
- safety precautions for performing maintenance (such as complete or partial shutdown)
- procedures for accessing subsystems or components to perform maintenance, including means of emergency shut-down

The information for use for complex systems should answer the following questions, for instance:
- Is it necessary (e.g.) to shut a subsystem down and select a different process path?
- Will a component to be worked on (or a component upstream or downstream) have to cool down or come to a stop?
- Will scaffolding be needed or will system components have to be disassembled?

When these questions have been clarified, the information for use may refer to the documentation of the respective component concerning the task. Such references should be as specific as possible (as a minimum, it should include a reference to the chapter of the supplier's information for use). It must be described how to re-integrate the component into the system upon completion of the task, e.g. how to perform a seal-tightness inspection, run-direction check and any required re-assumption of the standard process path etc.

With a view to availability of the information (efficiency!), a superordinate maintenance schedule should be issued, considering the maintenance cycles of all components.

Similarly, a superordinate summary for troubleshooting should be prepared, also containing information on interfaces. Additionally, a superordinate spare parts list and list of consumables, including information on their availability, should be prepared.

In many commercial systems the information on maintenance, troubleshooting, spare parts and consumables are part of the plant management system.

7.10.12 Troubleshooting and repair of the supported product by non-skilled and skilled persons

7.10.12.1 General

As outlined in CHAPTER 7.10.10, the need for target oriented writing (see CHAPTER 5.2.3 and CHAPTER 6.2.2) of information for troubleshooting and repair, according to the standard, requires providing separate information for skilled persons and non-skilled persons.

7.10.12.2 Information for non-skilled persons

According to the standard, the following information must be provided for non-skilled persons:
- methods of repair
- methods for testing after repair

The standard additionally suggests providing troubleshooting guidance or a list of frequently-asked questions (FAQs) with appropriate illustrations.

In this connection, all information needed to ensure safe use of the product upon repair must be provided.

7.10.12.3 Information for skilled persons

Information for skilled persons, as per the standard, includes the following :
- methods of repair
- methods for testing after repair

Additionally, the following information must be provided (if applicable):
- A list of indications for fault identification and its location for normal operation is required by the standard. Additionally, it may be useful to include a similar list for all other operating modes (setup operation, maintenance, fault etc.) in the information for use
- Descriptions of built-in diagnostic systems to aid fault detection, enabling the target audience to identify the cause of the fault as quickly as possible
- Directions for recording of alarm and trip conditions, e.g. in the shape of an "alarm and trip list", indicating the default and limit values
- Directions for shutting down and isolating of malfunctioning units and for starting standby or alternative systems, if applicable
- Instructions for troubleshooting and repair
- Contact details of the supplier or others from whom technical assistance can be obtained

For further notes refer to subsection 7.10.12.3 of the standard.

7.10.13 Replacement of parts by non-skilled and skilled persons

Information on replacement of parts by skilled or non-skilled persons must be provided separately, as per the standard, to meet the requirement for target oriented writing (see CHAPTER 5.2.3 and CHAPTER 6.2.2).

Generally, the following information must be supplied (specific for the respective target audiences):
- method of replacement
- method of inspection upon replacement (if needed)
- spare parts supply

If a refit of labels is necessary during the intended lifetime of the product, information about the locations of the labels and their replacement must be given, as per the standard.

7.10.14 Disassembly, recycling, disposal of the supported product

7.10.14.1 General

The standard requires that information for use includes information on disassembly, recycling or disposal of the product.

7.10.14.2 Disassembly

Disassembly work, which the target audience may carry out without incurring risk and while ensuring environmental protection, must be described.

7.10.14.3 Recycling

Where specific procedures are necessary for recycling of the product or its components, the standard requires that information for use includes information which is in accordance with the appropriate legal requirements and standards.

7.10.14.4 Disposal

As per the standard, information on disposal must be provided, taking into account safety considerations and environmental requirements.

For products containing data or intellectual property such as software, the standard requires that the supplier identifies the following in the information for use:
- Responsibilities concerning deletion/disposal
  Note that the requirements for disposal of data may change during the product life cycle.
- Determination of ownership of stored data, and of hardware and software assets
- Procedures for retention or destruction of data as well as confirmation to that effect

7.11 Safety-related information

7.11.1 Types of safety-related information

The standard differentiates three types of safety-related information:
- safety signs and product safety labels (see standards referenced by the standard: ISO 7010:2011 and ISO 3864, especially part 2)
- safety notes (see CHAPTER 7.11.4)
- warning messages (see CHAPTER 7.11.5)

The standard requires that safety-related information be included with the intended use (see CHAPTER 7.11, especially CHAPTER 7.11.4). The content requirements for such information are:
- the intended environment (e. g. outdoors, indoors, in a zone with potentially explosive atmosphere, under hot/cold conditions)
- the tools needed; additionally, information about the materials and the prerequisites with respect to the operating conditions should be included
- how to ensure that safety features of the product to be installed have been installed and are functioning
- the intended use
- the identification of permitted and inadmissible modifications of the product

For further notes refer to subsection 7.11.1 of the standard.

7.11.2 Location of safety-related information

According to the standard, safety notes must be provided at the beginning of the information for use in a separate clause or section. The clause/section must be clearly identified and must have a heading which emphasizes the importance of the contents. It is reasonable to place safety notes after the table of contents.

In step-by-step instructions, warning messages must be given in the context in which a hazard is likely to occur. It must be ensured that the warning messages are placed **before** the task. For presentation of warning messages see CHAPTER 7.11.5.

If personal protective equipment is necessary, the standard requires that the information to this effect is located at the beginning of the task description and is additionally placed on the package and the product, if applicable. Any note concerning personal protective equipment should be supported by use of pictograms (see ANSI Z535.4 and ISO 7010).

7.11.3 Precautions for particular target audiences

To satisfy the requirements for target orientation (see CHAPTER 5.2.3 and CHAPTER 6.2.2) safety-related information in particular shall be presented in a way that meets the needs of the target audiences. Therefore, the standard requires to place such information at the beginning and to highlight it for emphasis.

The standard mentions the following example for information targeted at particularly vulnerable audiences:
- children
- older persons
- persons with disabilities
- persons with electrically controlled medical devices, such as cardiac pacemakers
- persons who suffer from allergic reactions to specific chemicals
- persons who are sensitive to lighting effects (e.g. strobe lighting)

The list of particularly vulnerable target audiences contained in the standard is not exhaustive. A study of the target audiences, and whether these need to be considered particularly vulnerable under certain circumstances, should be done as part of the risk assessment; the findings must then be considered appropriately in the information for use.

Suitable warning messages must be provided, if the product could potentially affect electrically-controlled medical devices, such as cardiac pacemakers.

The following pictogram may be used to support such notes:

Figure 7: Pictogram "No persons with cardiac pacemakers", ISO 7010 P007

7.11.4 Safety notes

According to the standard, safety notes in information for use, as a minimum, must indicate the potential hazards. If possible, the causes of the hazards, the behaviour required to avoid the hazards and the probable consequences of not avoiding them should be stated.

At the outset of the safety notes a general statement shall highlight the importance of reading them.

Example:

To avoid possible harm, read and follow these instructions:

The subsequent safety notes must include information on the following:
- Identification of hazards with appropriate graphics, including:
  - ▹ Reasonably foreseeable misuse (see CHAPTER 7.1), including its consequences; for instance, use of a turning machine for machining of unsuitable (brittle) material, may cause small parts to be ejected
  - ▹ Inadmissible or inadvisable modifications of the product (see CHAPTER 7.11.1) However, before describing inadvisable modifications to the product, one should consider, as part of the risk assessment, if the idea to manipulate the

product would not be planted in the minds of the target audience by its being mentioned.
  - ▹ Emissions of the product, e. g. radiation, hazardous substances, noise
- Safety precautions, such as wearing of personal protective equipment, observance of safety signs etc.
- Notes concerning re-establishment of safe operation upon a fault or emergency situation

For further notes refer to subsection 7.11.4 of the standard.

Some of these requirements are redundant in the standard. The person preparing the information for use should carefully consider where to place such information as part of the design of the information product (see CHAPTER 6).

7.11.5 Warning messages

7.11.5.1 General

The standard requires that warning messages included in the information for use are concise. Where possible, the causes and sources of hazards, the measures to be taken to avoid the hazard, and the possible consequences of failure to avoid the hazard should be indicated, as is required regarding safety notes (see CHAPTER 7.11.4).

According to the standard, each warning message must be placed either before the first step or incorporated just before the relevant step of a step-by-step instruction.

Where avoiding a hazard requires following detailed safety procedures, these should be presented elsewhere in the information for use, but not in the warning message itself.

7.11.5.2 Design

To fulfil the standard, warning messages must be designed consistently and conspicuously. The following must be considered:
- Warning messages begin with a signal word (see CHAPTER 7.11.5.3 and CHAPTER 7.11.5.4)
- Limit the text and/or illustrations to the essentials
  Contents should include:
  - ▹ type and source of hazard (e. g. hot waffle iron)
  - ▹ consequence of failure to observe the following notes (e. g. burns)
  - ▹ notes on how to avoid the hazard (e. g. take food from the waffle iron with a fork, let waffle iron cool off before cleaning)
- Conspicuous design
  e. g. adapted to ANSI Z535.6. Warning messages should be designed to arouse attention, while at the same time not interrupting the reading flow. The effectiveness of warning messages should be scrutinized by usability testing (annex A of the standard, see CHAPTER 9.15).
- Make warning messages visible to all persons exposed to hazards
  In some cases this will require placing warning messages – or at least attention arousing pictograms – not only in the information for use but also on the product itself.

- Place warning messages as determined by the risk assessment
  This requires that the risk assessment contains information about the phase of life and the hazard location.

The four-elements method may be followed to design warning messages:
(1) Signal word
(2) Type and source of the hazard
(3) Consequences of failure to observe the warning
(4) Avoidance of hazard (measures needed to counteract or avoid the hazard)

7.11.5.3 Signal words for harm to persons

The standard recommends using the signal words from ISO 3864-2. These are defined there as follows, with a very close relation to the definitions in ANSI Z535.6:

Table 8: Signal word for harm to persons

DANGER	Signal word used to indicate an imminently hazardous situation which, if not avoided, will result in **death** or **serious injury**
WARNING	**Signal word used to indicate a potentially hazardous situation which, if not avoided, could result in death** or **serious injury**
CAUTION	Signal word used to indicate a potentially hazardous situation which, if not avoided, could result in **minor or moderate injury**

For further information refer to CHAPTER 12.5.

However, the standard also acknowledges that signal phrases such as "DANGER OF DEATH", "RISK OF BLINDING", or "BEWARE OF FUMES" may be more effective in certain circumstances.

The selection of the suitable signal word should be based on the reasoning in the risk assessment.

7.11.5.4 Signal words for damage to property

So far ISO has not defined any signal words for damage to property. ANSI Z535.6 suggests using "NOTICE". Protection from personal harm is of utmost priority in any setting. However, damage to a product may have substantial financial consequences on the one hand, while it cannot be excluded that a damaged product may subsequently affect persons on the other hand. Therefore, IEC/IEEE 82079-1 suggests using signal words such as "NOTICE", "ATTENTION" or "TAKE CARE" if there is a risk of such damages.

For the German language the signal word "ACHTUNG" (ATTENTION) has prevailed in many manuals.

7.11.6 Safety-related information in quick-start guides

The standard requires that quick-start guides include safety-critical information relevant to the tasks described in the quick-start guide. Moreover, similar requirements applying to any other type of information for use apply. The following applies additionally:

- The quick start guide must make clear to the target audiences why and when they need to read additional safety-related information in the information for use
- At the beginning, the quick-start guide should include a reference to the safety-related information in the information for use, which is made conspicuous and unambiguous by means of the general warning sign and/or a signal word. The standard references ISO 7010-W001:2011-05 "Graphic symbols – Safety colours and safety signs - Registered safety signs". The sign number W001 in ISO 7010 is the most well-known general warning sign:

Figure 8: General warning sign W001 to ISO 7010

- The text of the reference shall describe the possible consequences of not complying with the general safety messages.

7.12 Instructions for assembly of self-assembly products

Concerning instructions for assembly of self-assembly products the standard requires that these present the sequence of assembly actions in a step-by-step sequence of illustrations, each accompanied by text positioned adjacent to the step.

Where necessary, explanatory notes and warning messages (see CHAPTER 7.11.5) shall be provided.

As per the standard, the information for use must include the following for each assembly step:

- the components and other objects involved in that step, e. g. tools and consumables
- alignment and orientation of components to be assembled. The notes contained in the information for use should agree with orientation points on the actual components
- actions to be performed for assembly. Best by means of a step-by-step instruction, see subsection 8.3.4.1 of the standard
- safety precautions to take for the actions, warning messages, where needed (see CHAPTER 7.11.5)
- description of foreseeable errors that should be avoided
- verification of correct mating of components and fastening with the required strength (torque)

If the assembly of the product requires special skills, the standard requires including information to this effect in the instructions taking the form of a warning such as "for assembly by a qualified electrician" or "assembly by adults required".

If the risk assessment has revealed that foreseeable assembly errors may compromise the safety of the completed product or the safety of a person while undertaking the assembly, the standard requires addressing this in the risk management (see CHAPTER 6.2.7).

The standard recommends usability testing (see CHAPTER 6.3.3) to evaluate whether the notes included are sufficient.

7.13 Information for use for complex systems (plant documentation)

For complex systems, such as processing plants, the standard requires including the restrictions for use and safety notes in the instructions for use on the overall system level. For operational tasks (e.g. commissioning, decommissioning, further information see CHAPTER 7.10.5 and CHAPTER 7.10.11) step-by-step instructions for the system shall specify the users' tasks. The standard clearly shows that it is not sufficient to simply describe the interrelation of the system components.

IEC 81346-1 (successor of IEC 61346-1) presents one way of structuring descriptions of complex systems; it arranges them by the descriptions of the subsystems and components. IEC 61346-1 is the basis of identification systems used for power stations (e.g. German KKS or the RDS-PP).

7.14 Information security and data privacy

7.14.1 General

Today's technology with relative ease allows to prepare and provide information for use that is aligned to the needs of a specific target audience (see CHAPTER 5.2.3 and CHAPTER 6.2.2). Especially where information for use is provided in electronic format, this can be controlled effectively by requiring input of personal data.

To ensure data protection, the standard requires that such data is protected from misuse. Thus the information for use must include information on how users should act to ensure confidentiality, integrity, and accessibility of such data.

For remote controlled or net-attached products the information for use includes features to avoid misuse or unauthorized access (hacking), such as changing default passwords.

7.14.2 Functions for access control or protection of sensitive data

According to the standard, the most common methods of providing information protection or assurance must be taken into account when creating, exchanging, delivering or storing information (encryption, backup, data masking, and data erasure).The development of technology in this area is very rapid, requiring that the responsible persons keep abreast with the state of the art, or better still, the state of science and technology.

Guidelines for cyber security are available free of charge from the Bundesamt für Sicherheit in der Informationstechnik (German Federal Office for Information Security).

Additionally refer to the standard series ISO/IEC 27000 on information security. The standard ISO/IEC 27002 Information technology – Security techniques – Code of practice for information security controls for instance, includes recommendations on diverse monitoring mechanisms for information security, especially outlining measures against cyber attacks.

This topic should be addressed early on in the design and development process and the IT security department or external consultants should be involved.

7.15 Training

If training is required for the target audience(s) of a product, the standard requires calling attention to this fact in the information for use.

8 Structure of information for use

List of requirements of the standard for the structure of information for use

Subsection	Requirement to be checked	Notes
General requirements for the structure		
8.1 General	– The structure of the information for use enhances its user-friendliness and comprehensibility – Complex information for use is clearly divided into convenient parts serving specific purposes and is presented in a consistent format – Printed information for use comprising multiple documents (see CHAPTER 11.2.1) the information on the front covers or spines makes it easy to distinguish one document from another – If information for use is integrated from multiple sources, the following has been observed: ▹ the documents are consistently numbered ▹ methods of integration of information for use, for example, integration of a copy, adaptation of the content, cross-reference to the source document, or a specific section in the source document have been implemented ▹ the terminology and cross-references used are consistent	
8.2 Information types	The section does not contain requirements, only recommendations	
Structuring of step-by-step instructions		
8.3.4.1	– A step-by-step instruction contains preliminary information, the actual instructions, and completion information	
8.3.4.2	The preliminary information includes the following: – A brief overview of the purpose of the procedure and definitions or explanations of necessary concepts not included elsewhere – Identification of technical or administrative activities required before starting the task – A list of resources the target audience needs in order to complete the task. It can include, for instance: tools, other people (helpers), relevant data, documents, passwords (see CHAPTER 11.2.3), required or additional software or equipment – Relevant warnings, cautions, and notes that apply to the entire procedure	
8.3.4.3	– Instructional steps are numbered using Arabic numerals and presented in the order of performance – Relevant warning messages and cautions are provided before the applicable instructional steps – Each step provides a single action: ▹ Instructional steps indicate the expected result ▹ Error messages and recovery procedures which could occur are mentioned	

Subsection	Requirement to be checked	Notes
8.3.4.4	– The end of a set of step-by-step instructions is indicated, allowing the target audience to determine that the procedure has been completed successfully	
Navigation and information delivery		
8.4.1	– The information for use includes elements that make it easy to navigate the contents	
8.4.2.1	– Where more than two (printed) pages have been provided, the pages are numbered	
8.4.2.2	– If the volume exceeds 12 pages, a table of contents consistent with the contents is included	
8.4.3.3	– A mechanism for electronic search has been provided in information for use delivered electronically	

8.1 General

The standard requires that the structure of information for use must enhance the usability (user-friendliness) and comprehensibility.

One of the central requirements of the standard is that the information products must provide for efficient accessibility of information in any given application context. In case of power failure for instance, an emergency power supply should ensure that information on further action can still be displayed. Printed documents should be stored so that they can be located quickly – if possible directly on/at the product. To achieve this, a well though-out structure of the information for use is indispensable.

To ensure this, the standard provides general requirements and detailed requirements for the exterior and interior structure.

If printed information for use comprises multiple documents, the information on the front covers or spines shall make it easy to distinguish one document from another. To achieve that, it may be useful to subdivide it into parts that suit the application and/or target audience.

For information for use for complex systems (e. g. overall documentation of a processing plant), the standard allows using the information for use of the elements integrated into the system as parts of overall documentation (also see CHAPTER 10.2.1).

For general structuring of information for use, the following features may be used as part of a conventional setting:

- Table of contents

  The structure of the table of contents for products coming under legal regulations (e. g. EU directives) should follow the structure therein presented.

  Where the product comes under the Machinery Directive (2006/42/EC), refer to CHAPTER 12.4.

  The table of contents should not exceed four levels (see example in figure 23 in the annex).

- Marginal notes
  Remarks on the margin of the information for use.
  Explanations, keywords or summaries of the respective section may be placed here.
  Marginal notes may also appear in the table of contents to point out to the target audience the relevance of the section.

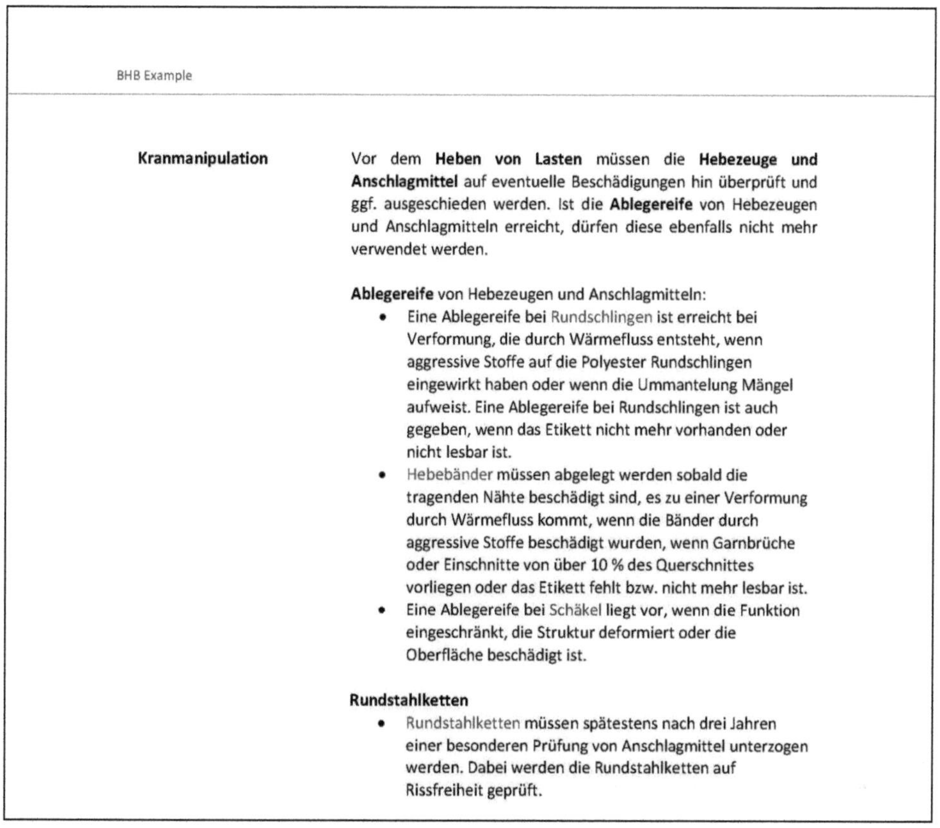

BHB Example

Kranmanipulation

Vor dem **Heben von Lasten** müssen die **Hebezeuge und Anschlagmittel** auf eventuelle Beschädigungen hin überprüft und ggf. ausgeschieden werden. Ist die **Ablegereife** von Hebezeugen und Anschlagmitteln erreicht, dürfen diese ebenfalls nicht mehr verwendet werden.

Ablegereife von Hebezeugen und Anschlagmitteln:

- Eine Ablegereife bei Rundschlingen ist erreicht bei Verformung, die durch Wärmefluss entsteht, wenn aggressive Stoffe auf die Polyester Rundschlingen eingewirkt haben oder wenn die Ummantelung Mängel aufweist. Eine Ablegereife bei Rundschlingen ist auch gegeben, wenn das Etikett nicht mehr vorhanden oder nicht lesbar ist.
- Hebebänder müssen abgelegt werden sobald die tragenden Nähte beschädigt sind, es zu einer Verformung durch Wärmefluss kommt, wenn die Bänder durch aggressive Stoffe beschädigt wurden, wenn Garnbrüche oder Einschnitte von über 10 % des Querschnittes vorliegen oder das Etikett fehlt bzw. nicht mehr lesbar ist.
- Eine Ablegereife bei Schäkel liegt vor, wenn die Funktion eingeschränkt, die Struktur deformiert oder die Oberfläche beschädigt ist.

Rundstahlketten

- Rundstahlketten müssen spätestens nach drei Jahren einer besonderen Prüfung von Anschlagmittel unterzogen werden. Dabei werden die Rundstahlketten auf Rissfreiheit geprüft.

Figure 9: Example of marginal notes

- (Living) column titles
  Remarks in the header and/or footer of a document.
  These may for instance contain information about the current chapter, the current revision, and the product supported by the information for use.

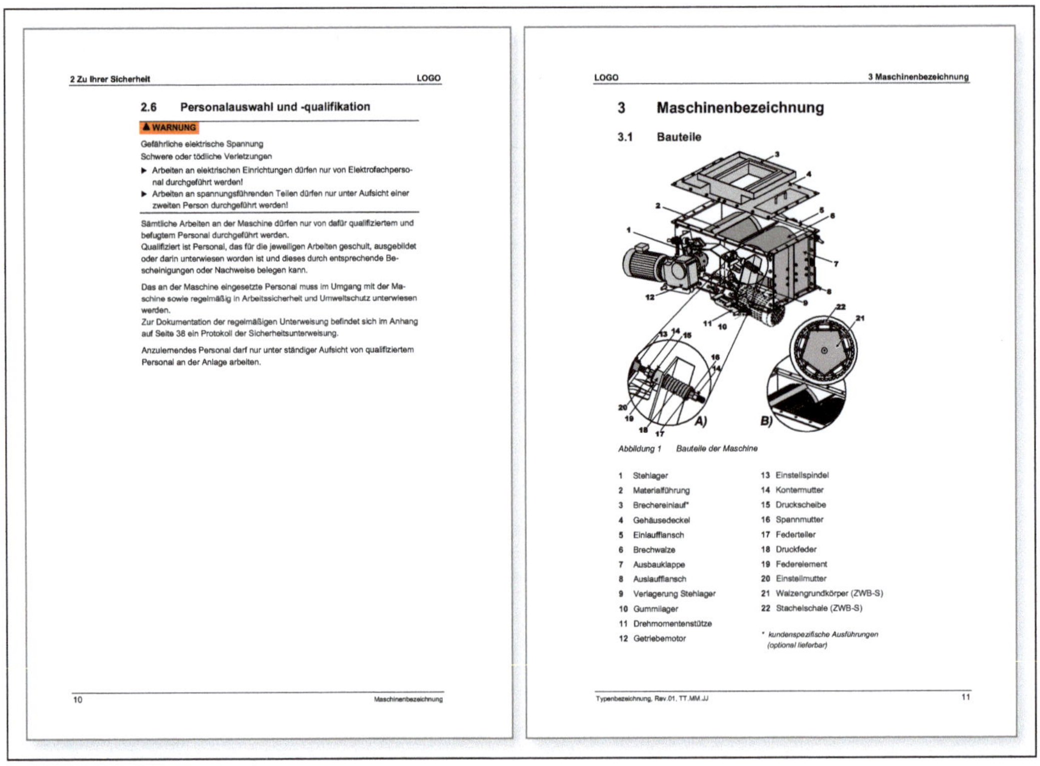
2 Zu Ihrer Sicherheit LOGO

2.6 Personalauswahl und -qualifikation

⚠ WARNUNG

Gefährliche elektrische Spannung

Schwere oder tödliche Verletzungen

- ▶ Arbeiten an elektrischen Einrichtungen dürfen nur von Elektrofachpersonal durchgeführt werden!
- ▶ Arbeiten an spannungsführenden Teilen dürfen nur unter Aufsicht einer zweiten Person durchgeführt werden!

Sämtliche Arbeiten an der Maschine dürfen nur von dafür qualifiziertem und befugtem Personal durchgeführt werden.
Qualifiziert ist Personal, das für die jeweiligen Arbeiten geschult, ausgebildet oder darin unterwiesen worden ist und dieses durch entsprechende Bescheinigungen oder Nachweise belegen kann.

Das an der Maschine eingesetzte Personal muss im Umgang mit der Maschine sowie regelmäßig in Arbeitssicherheit und Umweltschutz unterwiesen werden.
Zur Dokumentation der regelmäßigen Unterweisung befindet sich im Anhang auf Seite 38 ein Protokoll der Sicherheitsunterweisung.

Anzulernendes Personal darf nur unter ständiger Aufsicht von qualifiziertem Personal an der Anlage arbeiten.

10 Maschinenbezeichnung

LOGO 3 Maschinenbezeichnung

3 Maschinenbezeichnung

3.1 Bauteile

Abbildung 1 Bauteile der Maschine

1 Stehlager	13 Einstellspindel
2 Materialführung	14 Kontermutter
3 Brechereinlauf*	15 Druckscheibe
4 Gehäusedeckel	16 Spannmutter
5 Einlaufflansch	17 Federteller
6 Brechwalze	18 Druckfeder
7 Ausbauklappe	19 Federelement
8 Auslaufflansch	20 Einstellmutter
9 Verlagerung Stehlager	21 Walzengrundkörper (ZWB-S)
10 Gummilager	22 Stachelschale (ZWB-S)
11 Drehmomentenstütze	
12 Getriebemotor	* *kundenspezifische Ausführungen (optional lieferbar)*

Typenbezeichnung, Rev.01, TT.MM.JJ 11

Figure 10: Example of a (living) column title [Source: information for use of EME GmbH, Erkelenz]

- Front covers and spines
  of the information for use should enable unambiguous allocation of the information to the supported product. If multiple types of a product have been described, this should likewise be indicated. To show and enable recognizing the product, it may be useful to include an illustration of the supported product.
- Mark-up
  helps understand the application context (see Figure 11).

Procedural instructions are structured as follows in this manual:

➔ At the beginning of a procedure you are prompted to take action.

1. The steps needed for the action are numbered consecutively.

☑ This sign indicates information on the direct result of the action taken – this allows to check whether the step has been followed correctly.

Example of a procedural instruction:

➔ Replace the compression springs.

1. Shut the machine down as described on page 23.
2. Losen the lock nut (14, figure 3) and the clamping nut (16) on the adjusting spindle.

☑ The compression spring is now decompressed and can be removed.

Figure 11: Example of mark-up

Whether it is good and necessary to explain the mark-up at the beginning of information for use, is a subject of debate. Generally, the target audience will know what to expect in the manual. Also, considering the minimalism principle (see CHAPTER 5.3.3) it may not be necessary to explain the mark-up, because its consistent use in a specific application context will normally speak for itself.

However, it is useful to include mark-up rules in a guideline for authoring and editing.

In "mobile" information for use, context sensitive structures may prove useful. Use of taxonomies allows presenting the most suitable information for use to the allocated target audience at a given point.

When information from multiple sources is integrated into information for use, the standard says it is necessary to consider the following:

- consistent document numbering
- consistent methods of integration of information for use
- consistent terminology or use of cross-references

Consistent document numbering may follow the principles presented in EN IEC 61355-1. This is especially useful for complex documentation structures that occur in large-scale machinery systems.

The structure may follow the phase of life of the machinery system. The following example (see Figure 12) illustrates a customer-specific structure of information products from different sources, which are referenced in the overall information for use.

The safety notes are provided in a separate manual that is likewise used to instruct personnel. The electronic structure is also reflected by the structure of the physical ledgers.

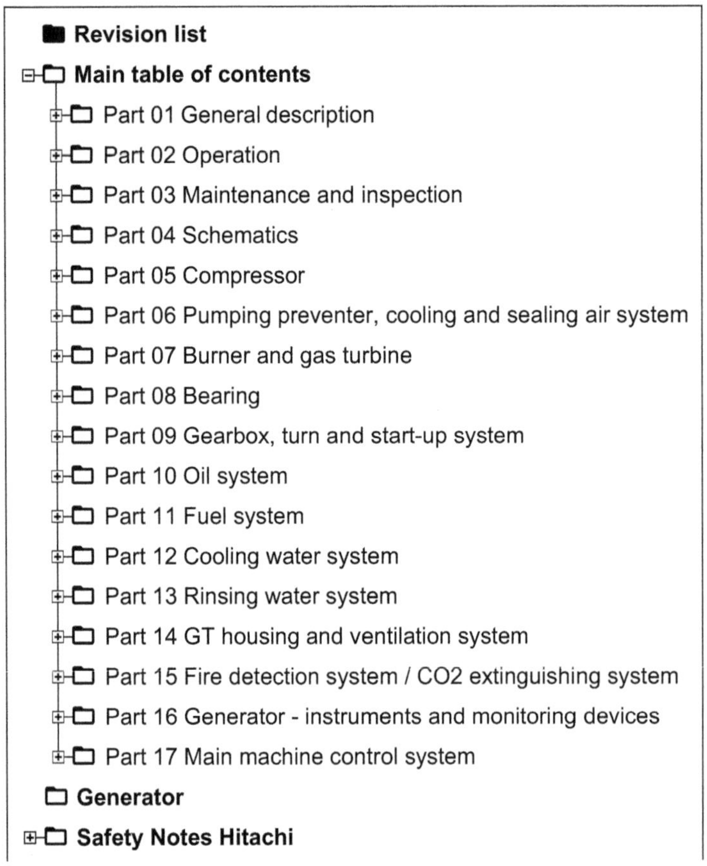

Figure 12: Example of the information for use for a complex machinery system

Note:
For further information on structuring of documentation see IEC 81346-1 and IEC 62023; for machinery see ISO 20607, section 5.

8.2 Information types

It is reasonable to place conceptual information at the beginning of a documentation. In it, the target audience may be informed of

- the conventions used in the information for use (e.g. how warning messages are structured and how cross-references are marked up),
- how the product is designed and
- what the pertinent controls of the product are.

Instructional information is meant to enable safe and efficient performance of tasks with the supported product. The supplier should determine how to address the target audience in the different settings (see CHAPTER 6.3.2).

Reference information is meant to help navigate inside and outside the information for use. It should be designed to enable unambiguous and rapid identification of the information sought. In electronic information for use, this requirement can be met by use of hyper-links. For a mere paper documentation, however, other concepts need to be worked out (e.g. direct cross-reference to a page).

8.3 Structuring

Information modelling differentiates micro and macro structures.

Micro structuring is determined on the level of the individual building blocks, e. g. module, node, fragment. The names of the building blocks vary between the available editorial systems (content management systems) or is determined by company-internal definitions.

Microstructuring defines items that cannot be set in an XML scheme: e. g. mark-up elements such as footer, graphic captions, keeping elements together, plural (for translation), singular (for translation), language-neutrality, or assignment of functional entities such as (un)numbered lists, procedural instructions. Rules for use of variables (and their inflection, if required) as well as for cross-references may also be defined.

One information modelling option is offered by DITA (**D**arwin **I**nformation **T**ype **A**rchitecture) already mentioned in CHAPTER 2.4 . Further modelling options are presented, for instance, in DocBook, S1000D, functional design, Information Mapping, and the class concept scheme.

When it comes to modelling of the macro structure, reusability of the building blocks is the primary concern. The modules must be small enough to be reusable, but big enough to make them systematically accessible. To enhance reusability, the modules must be combined with the meta data and, where applicable, the taxonomies defined by the company.

Subsections 8.3.1 through 8.3.3 contain notes on what is considered structuring by the standard. More detail is given as of subsection 8.3.4.1. It requires that step-by-step instructions contain preliminary information, instructional steps and completion information.

According to subsection 8.3.4.2 of the standard, the preliminary information must include:

- a brief overview of the purpose of the procedure and definitions or explanations of necessary concepts not included elsewhere
- identification of technical or administrative activities required before starting the task
- a list of resources the target audience needs in order to complete the task, which can include tools, other people, data, documents, passwords, or additional software or equipment
- relevant warnings, cautions, and notes that apply to the entire procedure (also see CHAPTER 7.11.5.3 and CHAPTER 7.11.5.4)

Structure:	Example:
Preconditions: `preconditions`	**Preconditions**:
• Target audience performing the job	• skilled electrician
• PPE required for the job	• safety gloves
• Machine status	• machine is shut down and secured against switching on
• Tools needed	• special super-key
Procedural instructions:	→ Proceed as follows to change settings of the assembly programme:
→ Target: `condition`	a) Select the assembly programme to be edited.
a) Step `action`	▸ A submenu opens.
▸ Intermediate result `intermediate result`	b) Set the number of cylinders and sets in a work cycle.
b) Step `action`	c) Set the cylinder programmes and positons in the WC.
▸ Result `reaction`	d) Save settings with the key F9.
	e) Quit the menu with the key F10.
	▸ The assembly programme is now set up.

Figure 13: Structure and example for a step-by-step instruction

According to the standard, the instructional steps must be numbered with Arabic numerals and must be presented in the order of performance (see subsection 8.3.4.3). In the example shown above (see Figure 13), Arabic numerals have been replaced by alphabetic numbering.

Relevant warning messages and cautions shall be provided before the applicable instructional steps.

Concerning formulation, the standard recommends using the imperative. Additionally instructions should not leave questions open that might cause the reader to interpret the meaning.

According to the standard, the end of a set of step-by-step instructions must be indicated, to allow the target audience to ascertain whether the procedure has been completed (see subsection 8.3.4.4). It is recommended to clearly indicate the end and important intermediate results of the procedure (see Figure 13, the results have been marked with green dots).

8.4 Navigation and information delivery

8.4.1 General

The standard differentiates between printed information for use and dynamically (electronically) delivered information for use.

For both forms of delivery it requires that information for use contains elements allowing easy navigation. For more information refer to CHAPTER 8.1.

8.4.2 Printed information for use

As already required in the previous version of the standard, only documents containing more than two pages must include page numbering (see subsection 8.4.2.1).

Note:
The additional indication of the total number of pages, following the pattern "page (n) of (m) pages", may be useful to enable checking for completeness.

A table of contents must be included, if the number of pages exceeds 12 pages, unless it can be demonstrated that it is not necessary. This is an amendment compared to the previous version of the standard, which required a table of contents if the number exceeded only four pages. The experts of JWG 16 have decided this based on their experience, considering the previous requirement exaggerated.

The requirement that the headings and page numbering shall be the same as in the text has remained. A new requirement is that a table of contents should have enough levels to help the target audience find information efficiently. Nonetheless, the table of contents should—if possible—not have more than four levels of headings. This new recommendation likewise reflects the experience of the experts in the committee. The recommendations concerning the index have not changed when compared to the previous version of the standard, but have only been extended by an informative reference to ISO 5963.That standard outlines methods for analysis of documents, to determine their content and then define terms to be indexed.

8.4.3 Electronic delivery

The standard contains only few requirements for electronic delivery, adding a number of recommendations.

The standard requires that electronically delivered information for use must include search functions, including, as a minimum, search for text strings. Additionally, the standard recommends that search should not only be based on full text search, but should be supported by search algorithms, taxonomies, thesauruses, and other available technologies. In subsection 9.6 the standard again requires that electronically delivered information for use be searchable. In this section, the standard additionally recommends including a feature allowing to display search results despite common misspellings.

Where individual topics are linked to other relevant topics, the linking shall not distract or demotivate the target audiences, e.g. a user should not be prompted to follow one link after the other and thereby lose track. If different media deal with the same topic (e.g. by mere text, video tutorials, or audio files), such contents should be cross-referenced or linked so that the information can be used together.

Where the standard speaks of individualized or context-sensitive delivery, it refers to something that is different from the common understanding of these terms. In the standard, individualized delivery means that only one particular piece of information needed to attain a specified goal, or perform a specific action is displayed. It is recommended that the full information is additionally delivered by a different method or in another form.

However, speaking about individualized delivery, the standard also refers to delivery of information to/for a specific target audience. This may refer to specialized personnel or a group of persons matching a certain user profile, for instance. As an example the standard mentions users that have paid for access. This example goes to show that dynamic information delivery may be considered and additional service (to be paid for).

In addition to individualized delivery, the standard explicitly mentions context-sensitive delivery. This refers to information for use that is delivered by an electronic device tracking additional information, such as geographical data, temperature, voice data or visual patterns, whereas such context information is then integrated into the information for use.

For instance, this may apply to a device that provides information for use and at the same time is capable of voice or face recognition and thus "knows", whether the person using the device is service personnel or an operator, and then bases the information delivery on what has been recognized. Use of such additional information of course depends on the consent of the person using such a device.

9 Media and format

According to the standard, the editorial style conventions for the format of information for use and for the media to be used shall be determined based on the type of target audience(s) and their needs. The media and formats must be selected so as to allow easy access to information for use throughout the intended lifetime of the product.

In addition to the classic printed format of information for use, the standard allows for other publication media. Above all, the selection of suitable media must conform with the legal requirements.

The media presented as examples in the standard can and should always be used as supplements (see CHAPTER 11.2.4).

In subsection 9.1 the standard contains a comprehensive list of examples of media that may be used to deliver information for use (for instance text, videos, AR and VR, brochures).

Also, the media mentioned may of course be used in parallel and redundantly. However, it is important to meet the expectations of the target audience concerning the medium employed.

The VDI Guideline VDI 4500 Part 6 "Technical documentation - Documentation process - Publishing" contains further information concerning the publication process.

Note:
Augmented Reality (AR). *The term refers to computer-aided extension of perceived realities. Reality is extended by computer-based additional information or virtual objects by means of superimposition/cross-fading. In technical documentation, for instance, assembly personnel may see a superimposed image of the next operating step in their field of vision. Warning messages and additional information (e.g. about a tool to be used) may also be displayed.*
Virtual Reality (VR) *refers to the generation of a computer-based interactive environment. VR is especially suitable for training purposes, as for instance in flight simulators or the simulation of maintenance tasks to be performed on very expensive and/or highly-sensitive products.*

Information provided via the Internet must additionally conform to the requirements of ISO/IEC 40500.
In selecting publication media, the following should be considered
- the legal requirements,
- the requisites of the product and
- the needs of the target audience

Often it makes sense, to contractually agree with the customer on the media to be used.

The standard contains numerous requirements for delivery of different media, which must be considered when the information for use is compiled.

List of requirements in the standard concerning publication media for information for use

Subsection	Requirement to be checked	Notes
9.1 General	– The media and format allow easy access to information for use throughout the intended lifetime of the product	
9.2 Durability	– It is guaranteed that information for use printed onto or moulded into the supported product remains legible throughout the intended lifetime of the supported product – Information for use given only on packaging or in materials accompanying the supported product is provided in durable form	
9.3 Audio-visual demonstrations	– The sequence of animations in electronic media follows the order of activities to be performed – Animation sequences are accompanied by synchronized spoken or written text – Advertising or promotional content is kept separate from the information for use	
9.4 Location and availability	– Information for use is given in one or more locations, e.g. on the packaging, as a manual, on websites etc. (see CHAPTER 5.3.8) – Information concerning installation, troubleshooting, repair, performing recovery actions, and contacting technical support is provided in a form that is available when the supported product is not operating (e.g. upon failure of power supply, display failure etc.)	
9.5 Printable information	The section does not contain requirements, only recommendations	
9.6 User interaction and search features	– Electronic media applications provide navigation systems (preferably tree-like) – Menus, buttons, hyper-links, breadcrumb trails, and bookmarks are provided to enable orientation	
9.7 Downloadable information for use	– Information for use that is intended to be downloaded is provided in a form that allows for display on commonly used devices and operating systems – Websites (or other platforms, e.g. cloud storage) from which information for use can be downloaded meet the following requirements: ▹ they provide either the software needed, or a link for downloading the required viewing software; and ▹ they allow downloading at any time	
9.8 Suitability for the conditions of use	– The media employed are suitable for the conditions in which they are used	
9.9 Consistency of format	– Consistent formats are used throughout, for the user interface, data elements, field names, functions, pages, topics, and processes	
Legibility, readability and comprehensibility		
9.10.1 Text font sizes	– Text, safety signs and graphical symbols are clearly legible for the target audience(s)	

Subsection	Requirement to be checked	Notes
9.10.2 Contrast	– For printed, embossed or engraved information for use, the brightness contrast is as high as possible – The contrast requirements for text displayed on visual display terminals meets what is defined in ISO 9241-300	
9.10.3 Location on the product or its packaging	– The location of the information for use is such that it is easily legible for the user upon initial encounter	
9.10.4 Minimum heights	The section does not contain requirements, only recommendations	
9.10.5 Simple wording	The section does not contain requirements, only recommendations	
9.10.6 Function of information sections	The section does not contain requirements, only recommendations	
Use of visualization		
9.11.1 Graphical symbols and safety signs	– Graphical symbols in diagrams shall conform to IEC 60617 and ISO 14617, if these standards are applicable – Symbols in flowcharts conform to ISO 5807 – If a symbol registered in ISO 7000, ISO 7010 or IEC 60417 is applicable, the registered graphical symbol is used – Registered safety signs from ISO 7010 are used in information for use – Safety signs comply with the design principles presented in the standard series ISO 3864 – Safety colours specified in ISO 3864 are used	
9.11.2 Illustrations	– Text and illustrations follow the sequence of operations – Illustrations are placed as close as possible to the text to which they relate, so that they can be viewed adjacent to the relevant text (text-illustration combinations) – Explanatory text is placed directly adjacent to the illustration – The sequence of illustrations is logical and comprehensible	
9.11.3 Information content of illustrations	The section does not contain requirements, only recommendations	
9.11.4 Captions	– The relationship between the captions and the illustration is clear	
9.12 Tables	– Tables are set out clearly, informatively, and in a consistent design – Tables are positioned next to the relevant text, except for extensive reference tables, which may be placed in annexes	
9.13 Colours	The section does not contain requirements, only recommendations	
9.14 Icons	The section does not contain requirements, only recommendations	

Subsection	Requirement to be checked	Notes
Attracting attention to safety-related information		
9.15.1 General	– Safety-related information is highlighted by a signal word (see CHAPTER 7.11.5.3 and CHAPTER 7.11.5.4) – Warning messages about hazards or restrictions on use are displayed at least as prominently as the other information in documents – Safety notes shall be placed in the information for use in a prominent position before the use of the supported product is addressed and they are emphasized by a headline – Warning messages displayed on a screen are visible for the whole time where they are relevant – Illustrations intended to convey safety-related information uses colours, shapes, positioning or other means to make the information conspicuous	
9.15.2 Durability and visibility	– The visibility of warning messages is guaranteed for the entire intended lifetime of the supported product – When colours are used in warning messages, discolouration and fading are avoided	

9.1 General

The standard requires that the media and formats allow easy access to information for use throughout the intended lifetime of the product.

For products with long service life this requirement is hard to meet at the time the product is supplied. To achieve it, later distribution of information for use and the selection of publication channels must be prepared carefully and comprehensively. Diverse technical options are available for the publication channels, however, they need to be suitable for the respective target audience. The options for accessibility must match the product and ensure simple, permanent access.

The following examples give an idea of how information for use may be provided in addition to the documents that accompany the actual shipment of the product:

- QR codes or barcodes (see Figure 14) in the accompanying documents or on the product allow to download information for use via a web-link e. g. as a PDF/A file.
  However, it cannot be guaranteed that such technology will function in a more/less distant future.

  www.tekom.de

  Figure 14: Example of a QR code and a barcode

- Additionally, it must be ensured that the codes provided remain legible and permanently connected to the source.

Note:
Consider a possible expiry date of QR codes or barcodes generated.

- Furnishing the product with an NFC chip (Near Field Communication). Such chips may store (currently only short) information for use, which can be called up e.g. using an NFC compatible smart phone. The chip may also provide a link to the current information for use on the Internet.
  It must be ensured that the chips provided are legible and permanently connected to the source. Interference from the environment of the product used (e.g. strong magnetic fields) must be considered.
- Indication of the hotline on the product (e.g. by means of a sticker), via which the information for use may be requested by telephone or by e-mail. It will be practical to place an unambiguous identification of the product (information on the nameplate or individual product ID) next to the hotline information on the product. This method should be preferred, if the target audience is expected not to be (very) technophile.

9.2 Durability of chosen media

To fulfil the standard, the supplier must guarantee that the information for use remains legible throughout the intended lifetime of the supported product. Additionally, information given only on packaging or in materials accompanying the product must be in durable form. A note such as "Keep this information for reference" is not sufficient. Depending on the product, one should research the possibilities for provision of additional benefits to the target audience. Information for use for a microwave oven, for instance, could contain recipes. Or the packaging could be designed to form a practical storage box.

The requirement for durability of the information for use does not only refer to the legibility of the information (e.g. by retaining contrast), but also to the medium used for presentation of the information for use (e.g. CD, data goggles), it too must remain usable for the expected service life. Also refer to the notes on the subject in CHAPTER 9.1.

Note:
Technical developments are hard to foresee. Therefore, the selection of media regarding durability should follow recognized, commonly used, and, if applicable, standardized archiving methods. Experts for long-term data storage and retrieval of data may aid in selecting appropriate electronic media.

9.3 Use of animations or audio-visual demonstrations

If animations or audio-visual presentations are usedF, the standard requires that the information should follow the order of the activities. The animation sequences must be accompanied by synchronized spoken and/or written text. Additionally, information for use must be kept separate from promotional content.

9.4 Location and availability

According to the standard, the information for use must be available from one or several locations, depending on the needs. It must be considered that information for some activities may have to be available during standstill or failure of the product.

Indicators and displays, or controllers of the product, which hold information for use available, may be supplied with energy from an uninterrupted power supply (UPS) or an additional battery during a (short-term) standstill/failure. If the standstill (or failure of the product) may continue for a longer period, the information for use should additionally be available on a medium that is independent of the product itself (see CHAPTER 8.1.).

9.5 Printable information

The standard recommends ensuring the following when preparing printable information

- legibility of the information considering the selected format (also refer to table in subsection 9.10.1 of the standard),
- the volume (number of pages) of the printed document required to deliver complete information.

Already during the development of printable information for use, the supplier should keep an eye on preserving the legibility of its content. For further information concerning the information management process refer to CHAPTER 6. The same applies to electronically supplied information for use that is printable, e. g. PDF files.

9.6 User interaction and search features

To fulfil the standard, electronic media applications must take advantage of navigation systems. A tree-like structure is to be preferred. According to the standard, menus, buttons, hyper-links, breadcrumb trails, and bookmarks can be used to enable orientation within information for use.

Note:
Breadcrumb trails are a design pattern for presentation of graphic user interfaces. A breadcrumb may be represented by a line of text, which provides information about the branch of the application's content that the user is currently viewing.

As already mentioned within the frame of subsection 8.4, the standard requires that electronic information must be searchable; results should also be shown in case of common misspellings.

9.7 Downloadable information for use

Downloadable information for use should, where possible, allow display without the user having to download additional viewing software. This may be ensured, for instance, by integrating a suitable viewing programme into the website. Another option is displaying the information for use on a separate website. However, the information should also always be supplied to the target audience in a form that allows for offline use.

Further information can be found in ISO/IEC/IEEE 23026.

9.8 Suitability for the conditions of use of the supported product

The standard requires that the media employed are suitable for the conditions in which they are used.

To meet this requirement, a target audience analysis (see ISO/IEC 26514) and an analysis of the probable location of use of the product and/or the information for use must be performed. Additionally, the requirements from CHAPTER 9.2 and CHAPTER 9.4 must be considered.

Table 3 of the standard contains examples of unsuitable media. The following table lists examples of suitable media following the statements of the standard, but turning them into the positive:

Table 9: Examples of suitable media for different conditions of use

Conditions for use of the supported product	Suitable media for use of certain information
Low-light conditions	Illuminated text (e.g. reader with back-lighting), Virtual Reality (VR)
Noisy environment Use of hearing protection	Text and/or illustrations, pictograms
Wetness, humidity Clean-room conditions	Augmented Reality (AR), laminated paper manual, audio guides

9.9 Consistency of format

The standard requires consistent use of formats for user interfaces, data elements, field names, functions, pages, topics, and processes.

9.10 Legibility, readability and comprehensibility

9.10.1 Text font sizes and heights of safety signs and graphical symbols

Text fonts, safety signs and graphical symbols shall be clearly legible for the user from a comfortable distance, whether printed, integrated with the supported product, or computerized.

Table 4 of the standard lists the recommended minimum text font sizes and heights. This table differs very little from table 2 in the previous version of IEC 82079-1. The note on the subject likewise has not changed: The best legibility is achieved when optimizing the relationship between variables, including type font, font size, contrast, number of characters per line and line spacing. The revised standard additionally states that 70 characters per line and a line spacing of 120 % of the text font size should be selected for Latin characters. The standard requires that bold lettering, underlining etc. must not impair legibility. Target audiences with hearing or sight impairments must be taken into consideration when designing text, safety signs or graphical symbols.

9.10.2 Maximum brightness contrast

To fulfil the standard, text fonts, safety signs and graphical symbols must be designed readily readable for the target audience. For printed, embossed or engraved information for use, the brightness contrast must be as high as possible. The contrast requirements for text displayed on visual display terminals must meet what is defined in ISO 9241-300

Notes on the requirements from CHAPTER 9.10.1 and CHAPTER 9.10.2 are contained in table 4 of the standard.

9.10.3 Information for use provided on the supported product or its packaging

The standard requires that the location of on-product information for use and the angle between its surface and the vertical plane must be such that it is easily legible from the intended position of the user.

Where the information for use is meant for persons with particular needs, the positioning and location of the information for use shall consider these needs.

In course of the target audience analysis, it will be helpful to consider how the target audience will likely use the product and whether certain positions for the user/users have been defined. For instance, if the information for use is supplied in a box located on the product, it is reasonable to select a colour for the box that contrasts with the base colour of the machine and to attach the pictogram “Consult manual” from ISO 7010, M002, to the box:

Figure 15: Pictogram “Consult manual” to ISO 7010

Note:
Generally, when calling attention to the need to consult the manual, the internationally standardized mandatory sign should be used. Avoid using proprietary creations (see Figure 16).

	Please read this instruction manual carefully, especially the safety warnings marked with the symbol.
	Read this manual carefully. Make sure that you are familiar with the different controls, settings and handles of the equipment.

Figure 16: Proprietary note for use of the manual (avoid)

9.10.4 Minimum heights of safety signs and graphical symbols

Like its predecessor, the standard recommends considering the minimum heights for safety signs and graphical symbols listed in Table 4. Text font sizes, which are smaller than what is specified in the table, e. g. because of the size of the product, should only be selected upon carefully considering the possible safety consequences for the target audience.

Note:
In individual cases, country-specific legal regulations may require minimum heights differing from what is specified in the standard (e. g. individual acts of American Consumer Product Safety Councils [CPSC]).

9.10.5 Rules for simple wording

The standard recommends writing short sentences following the principle of "one expression, one simple meaning". The wording should be simple and clear. Colloquial expressions and uncommon regional variations of names and product features should be avoided, because this may not only impair the reception of information and its comprehensibility, but may also cause translation errors.

A manual should be understandable without a glossary, where possible. However, the target audience's needs must be considered. Trained technicians may be expected to have a greater vocabulary than end users. Use of consistent terminology is a prerequisite to meeting this requirement.

For systematic selection of the desired wording, a coarse classification into four main target audience groups may prove helpful:

- Disinterested layperson: Has little technical understanding and is unwilling to gain more. Manuals must be kept simple and understandable, use of technical terms must be avoided.

- Interested layperson: Has little technical understanding, but is willing to gain some technical understanding. Technical terms should be avoided, simple wording should enhance understanding.
- Practical user: Uses the product without reading the information for use carefully. Simple wording should increase the inclination to read. Technical terms may be used sparingly, must be defined in a glossary, however.
- Professional: Understands the technology and is capable of dealing successfully with demanding wording (e. g. medical personnel). Technical terms shall be used.

9.10.6 Function of information sections

The standard recommends making the function of information sections readily identifiable, allowing unambiguous identification of the different information types.

The identification of information sections may also be achieved by headings, marginal notes, and (living) column titles (see CHAPTER 8.1).

9.11 Use of visualization

9.11.1 Graphical symbols and safety signs

The standard recommends using graphical symbols in diagrams that conform to IEC 60617 and ISO 146117. The prerequisite for fulfilment of the standard, however, is, that symbols in flow diagrams conform to ISO 5807.

If a graphical symbol is registered in ISO 7000 or IEC 60417, the symbol must be used.

Graphical symbols and safety signs are a language-independent medium for submission of safety-related information. Therefore, only safety signs to ISO 7010 must be used in information for use, which have been designed according to the requirements of ISO 3864-x (including shapes and colours). Following this requirement guarantees that language independent safety-related information is understood by as wide as possible a target audience without use of additional information. For instructions to be used in the USA, the requirements of ANSI Z535 should take precedence.

The safety colours defined in the ISO 3864-1 standard series must be used exclusively for safety signs. The standard calls attention to the fact that rules for drawing specific diagrams are given in IEC 61082-1, ISO 5807, and ISO 15519-1. Use of standardized graphical symbols is also useful, because of the greater recognition factor of such symbols, not only to fulfil the requirements of the standard.

9.11.2 Illustrations

Illustrations should be self-explanatory and designed to focus attention on important details. They should be suitable for the target audience in the target cultures (localization of illustrations) and should be designed and selected individually for each purpose by competent graphic artists or technical illustrators.

Print quality or screen resolutions should be adapted to encourage rapid recognition.

The standard requires that illustrations showing an operation must follow the operational sequence.

Illustrations must be placed as close as possible to the text or operational sequence to which they relate (text-illustration combinations, see Figure 17). Where illustrations need explanatory text, such text must be placed directly next to the illustration. The sequence of illustrations must be logical and comprehensible (see Figure 18). According to the standard, rapid recognition of illustrations requires good print quality (exceeding 300 dpi) and sufficient screen resolution.

To fulfil this requirement, one might use a two-column page layout and/or a layout with two pages next to another (depending on the size of the illustrations). This will allow users to catch more information at a glance. If the illustration is too large to fit in one column of a two-column layout, the illustration may be placed on the even-numbered page and the related description on the opposite odd-numbered page. If several such two-page combinations are needed, a specific template for this should be developed.

Handlung 6.1:
Kontrolle der Fettfüllung der Hochdruckschmierbüchse.
Für einen störungsfreien Betrieb und eine lange Lebensdauer muss der Innenraum im Stehlager ständig mit Fett gefüllt sein. Die Hochdruckschmierbüchse gewährleistet eine permanente Fettzufuhr.

Wer: geschultes Bedienpersonal

Wann: monatlich

Wie: Durch das Schauglas die Fettfüllung der Schmierbüchse kontrollieren.

Ist die Füllung an der unteren Markierung (leer), Hochdruckschmierbüchse auffüllen, wie in Handlung 6.8 beschrieben.

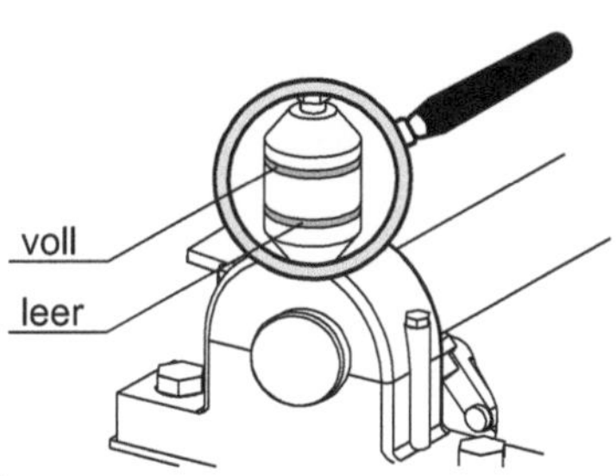

Abbildung: Hochdruckschmierbüchse

Figure 17: Example of a text-illustration combination (high-pressure lubrication cartridge)

Betriebsanleitung
Umbauständer

3.3 Lager wechseln / Rolle umbauen

1. Öffnen Sie die Druckluftversorgung für die luftbetriebene Hydraulikpumpe.

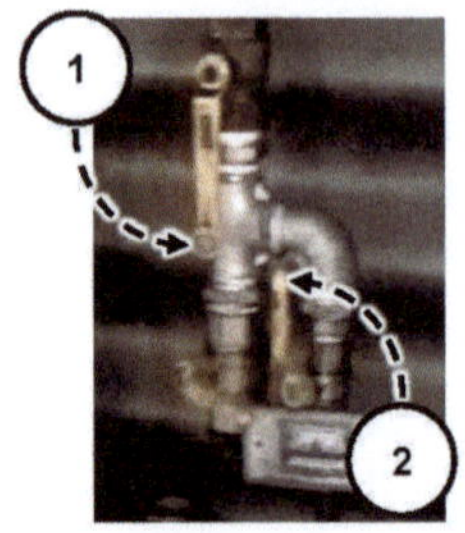

2. Öffnen Sie die Fixierklemme, damit Sie die umzubauende Rolle einlegen können.

3. Legen Sie die Rollen mit Hilfe des Hallenkrans in den Umbauständer ein.

SICHERHEITSANWEISUNG
Vorsicht beim Krantransport! Gefahr durch schwebende Lasten, quetschen und stoßen!

4. Schieben Sie den verstellbaren mobilen Werktisch unter das abzubauende Lagergehäuse.

5. Entfernen Sie die Lagerschalen durch lösen der Schraubverbindungen.

6. Entfernen Sie die untere Hälfte durch Absenken des Werktisches.
7. Nehmen Sie die obere Hälfte ab und legen Sie diese auf dem Werktisch ab.
8. Entfernen Sie die weiteren Komponenten:
   - Simmeringe
   - O-Ringe

Datum	Ersteller	Datum	Prüfer	ZDM-Nr.	Typ	Blatt	Version
01.01.20xx		15.01.20xx		**123456789**	**BHB**	23 von 31	**a**

Figure 18: Example of a text-illustration combination (operating instructions)

9.11.3 One illustration, one information provided

The standard requires that information overload in illustrations must be avoided. It recommends that illustrations should provide only information relevant to describing the related function. Illustrations, or detailed parts of these, should be repeated in the relevant parts of the information for use as needed.

9.11.4 Illustrations with captions

The standard recommends providing captions in illustrations. It requires that the relationship between the illustration(s) and the captions must be unambiguous. Captions should be placed in the same position in all illustrations (e. g. from top or bottom always).

9.12 Use of tables

The standard requires that tables be set out clearly. They must be informative and consistently designed. If descriptive text makes reference to a table (or vice versa), the text shall be placed adjacent to the table. Extensive reference tables should be placed in the annex, says the standard.

9.13 Use of colours

The standard suggests using colours as a supplementary means of drawing attention to particular sections of information only, if one must expect that the target audience includes people with colour-deficient vision.

Target audiences that definitely exclude persons with colour-deficient vision will be relatively rare. However, for some groups of professionals, notably pilots and electricians, colour-deficient vision can be excluded. But normally, one will have to expect that the target audience will include persons with colour-deficient vision.

Contrasting colours used in information for use should remain distinct upon printing on a monochrome printer. It is useful to consider these prerequisites during the target audience analysis and while the publication concept is developed.

9.14 Use of icons

The standard recommends using the same icon for a represented object throughout the information. If an icon represents changes (e. g. Δ for difference), it shall be used consistently. The standard also references IEC 62744, providing information on how to use icons to present information on the behaviour of a supported product.

9.15 Attracting attention to safety-related information

9.15.1 General

The standard requires that safety-related information is emphasized by an initial signal word (see subsections 7.11.5.3 and 7.11.5.4) in larger and/or different type font or size, by use of colours, safety signs or graphical symbols, lines, indents, or other means.

Warning messages about hazards or restrictions on use (e.g. "Not suitable for children under three years") are crucial for safety and must be displayed at least as prominently as the other information in documents provided with the supported product.

If warning messages are integrated between the steps of a procedure, the formatting of the warning message should not distract the target audience from reading the procedural information.

Safety notes must be emphasized by a headline that identifies the subsequent material as safety-related (in addition to the requirements in the standard, use of the signal word for safety instructions from ANSI Z535 may prove useful). In the information for use, they must be placed in a prominent position before the use of the supported product is addressed. In printed information for use, safety notes must be presented in the front part and before information for transport, assembly, operation, etc., is given.

Warning messages displayed on a screen must be visible as long as the signal triggering the warning is present. To fulfil this requirement, the standard recommends requiring the user to acknowledge the message before it is removed from view.

Illustrations included in safety-related information must use colours, shapes, positioning or other means to make the information conspicuous. Safety signs and graphical symbols should be used to emphasize key messages (see CHAPTER 9.11.1).

Proper risk assessment (to EN ISO 12100) is a key prerequisite for determining the correct signal words and researching all safety-related information.

A risk assessment will contain much information that the writer may base "his" information for use on or which he may even take over. Additionally, a risk assessment should allocate safety-related information to the information for use to be prepared.

The following list provides an overview of the type of information that may be relevant:

- Limits of the product (see CHAPTER 12.2)
- Phases of the life of the product (see CHAPTER 12.3)
- Location and type of guards and safeguards (including information on safe installation, disassembly, and functional inspection prior to commissioning)
- Intended use and reasonably foreseeable misuse
- General behaviour of personnel (information to be included in safety notes or the safety clause):
  - ▹ requirements for personal protective equipment (general)
  - ▹ requirements for and minimal scope of regular training
  - ▹ check-up of the pictograms attached to the product
- Behaviour of the personnel during hazardous tasks (information to be covered by warning messages or in the safety clause):

  - ▹ requirements for the specific personal protective equipment to be used for the task
  - ▹ minimum qualification for performing the required steps
  - ▹ warning messages following the four-stage principle:
    (1) Signal word
    (2) Type and source of hazard
    (3) Consequences of failure to observe the warning
    (4) Avoidance of the hazard (measures needed to counteract or avoid the hazard)
- requirements for education/training of the personnel (for the different tasks)

The residual risks must be clearly identified in a proper risk assessment report and must be marked so the writer of the information for use may filter out the required information.

To achieve this, a proper risk assessment must include especially the following information in concrete terms:
- hazard zones
- phase(s) of life, in which the hazard exists
- tasks/actions that involve (residual) risks
- classification of the hazard

To enable the writer of the information for use to select the appropriate signal word, the classification should be based on a standardized schematic.

One option for determining the signal word is the "Signal Word Selection Process" from ANSI Z535.6, which is also the basis for ISO 3864-2 (see Figure 19).

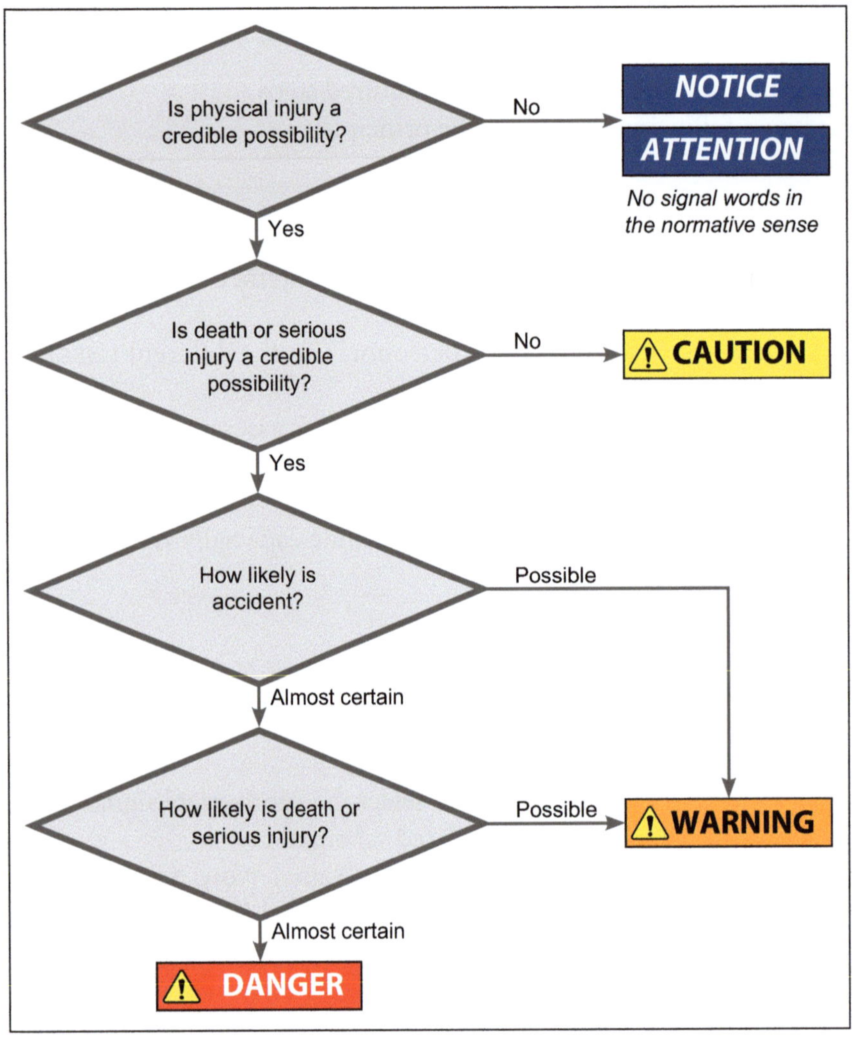

Figure 19: Selection of the signal word

Note:
ISO 3864-2, in the informative annex B, also contains an overview of the signal words in 19 different languages.

Resulting from the points previously mentioned, warnings of residual risks need to be presented in information for use with a signal word representing the respective risk level:

- DANGER indicates a hazard with a high level of risk which, if not avoided, **will result** in **death or serious injury**.
- WARNING indicates a hazard with a medium level of risk which, if not avoided **could result** in **death or serious injury**.
- CAUTION indicates a hazard with a low level of risk which, if not avoided, **could result** in **minor or moderate injury.**

In practical application the signal words will have to be chosen on the basis of the risk graph used during risk assessment. The annex uses different risk graphs to illustrate how a consistent evaluation scheme for determining the appropriate signal words may be drawn up (see CHAPTER 12.5).

9.15.2 Durability and visibility

To fulfil the standard, visibility of warning messages must be ensured for the intended lifetime of the supported product. When colours are used in warning messages, discolouration and fading must be avoided..

To meet this requirement, the concept for design of printed information for use should address the question of storage of information for use (e.g. in wet rooms, in industrial environments with strong contamination or in locations with direct sunlight). If this raises concerns, the target audience should be given the opportunity to reprint the information for use.

10 Professional competencies

As a novelty, the standard requires that preparation of information products must be assigned to competent persons. That is, the personnel involved must have pertinent professional qualifications.

This is defined in more detail by means of recommendations concerning the required professional competencies.

List of requirements in the standard concerning competencies of persons preparing information for use

Subsection	Requirement to be checked	Notes
10.1 General	– Information for use is prepared by competent persons	
10.2 Competencies	– To maximize the comprehensibility and safe use of a product, tasks related to information for use are assigned to competent persons	
10.3 Levels of proficiency	The section does not contain requirements, only recommendations	
10.4 Translators	– When information for use is translated from the original language into a target language, competent translators or language specialists are involved	

10.1 General

The experts of JWG 16 deemed it important to clarify that the preparation of information for use belongs in the hands of competent persons. Therefore, it was purposely formulated as a requirement.

Ideally, an organization should evaluate the tasks and define the competencies needed along the process for preparation of information for use and then assign the responsibility for the tasks to sufficiently competent persons.

The experts of JWG 16 clearly understood that the degree of fulfilment of this requirement will depend on the size of the organization, and that the full responsibility might be in the hands of one person in small organizations, while sizeable teams or a dedicated department may handle the preparation of information for use in larger organizations.

The standard differentiates task-related competencies and proficiency levels.

10.2 Task-related competencies

The standard emphasizes that to ensure the comprehensibility of information for use and safe use of the supported product, the roles within the process must be assigned to competent persons.

The standard, in all, outlines thirteen task-related competencies. It must be emphasized, however, that not everyone must have all of these competencies, but only those that are required for the tasks handled by him/her. The task-related competencies listed in the standard are aligned with tekom's competency model, which subdivides competency into seven fields of expertise.

tekom competency model with seven fields of expertise	Task-related competencies to IEC/IEEE 82079-1, referenced by means of the respective letter
1. Field analysis	**a)** Analysis of – the product life cycle – the target audience – the required information – the languages needed
2. Planning	**b)** Development of standardized processes for preparation of information for use **d)** Implementation of a successful support system, e.g. for content management and information management **m)** Fulfilment of leadership roles, preparation of project plans, management of service providers, communication with developers and suppliers, but also with customer-facing employees in the organization
3. Concept development	**c)** Development of content and media concepts as well as definition of appropriate content structure templates and styles **h)** Use of appropriate methodologies and tools for more efficient and effective design of the information management process
4. Content creation	**e)** Gathering of the required safety-related information from the development and design departments This especially refers to the ability to read and interpret risk assessments, extract the information relevant for the information for use, and then place them in the appropriate locations therein. **f)** Development of written and audio-visual content, including safety-related information **g)** Proofreading and correction of information for use **i)** Organizing the localization and translation of information for use **j)** Approval and release of information for use based on formal and functional quality and usability standards and criteria
5. Media production	**k)** Production of output media (that previously have been defined on the basis of the expectations and needs of the target audience and the use cases) and software-based instructions
6. Publication and distribution	**k)** Delivery of the information for use in the form of the output media prepared
7. Observation (monitoring) of the information product	**l)** Analyse the way in which the target audience benefits from the information for use. Appropriate methods and tools include website statistics, empirical testing and social media feedback. Based on such analyses, the writer of information for use is capable of further developing and continuously optimizing the information for use.

For more information on the tekom competency model scan the QR code with your smart phone or call up: https://www.technical-communication.org/fileadmin/tekom.de/user_upload/tekom_2018-04-18_Competence-Framework-Teaching-Perspective_V1_en.pdf.

The tekom competency model helps managers and human resources departments in working out job profiles and roles; it also supprts them in preparing planned improvement of the competencies of their employees. With the aid of profiling tools based on the competency model personal competency profiles can easily be drafted.

These competencies are imparted for instance by universities, professional training agencies and many publications and events organized by tekom (www.tekom.de).

10.3 Levels of proficiency

The standard suggests three levels of proficiency (see Figure 20):

Proficiency level 1 is the basic stage, encompassing a wide range of theoretical and factual knowledge in the area of work. It includes a set of cognitive and practical skills to find solutions to specific problems, and the ability to accept responsibility for accomplishing work. Additionally, the personnel should be capable of adapting their own behaviour to suit the relevant circumstances when solving problems. Proficiency level 1 could also be referred to as the "entry level".

The proficiency level 2 in addition to the above mentioned skills, first of all requires the ability to determine the requirements for the information products, taking into account the purpose and domain to which they apply. It may also be referred to as the "advanced level".

The proficiency level 3 additionally requires the ability to manage the process of information creation, develop content strategies, take the responsibility for conformity to applicable standards and policies, and lead information development teams. It may also be referred to as the "management level".

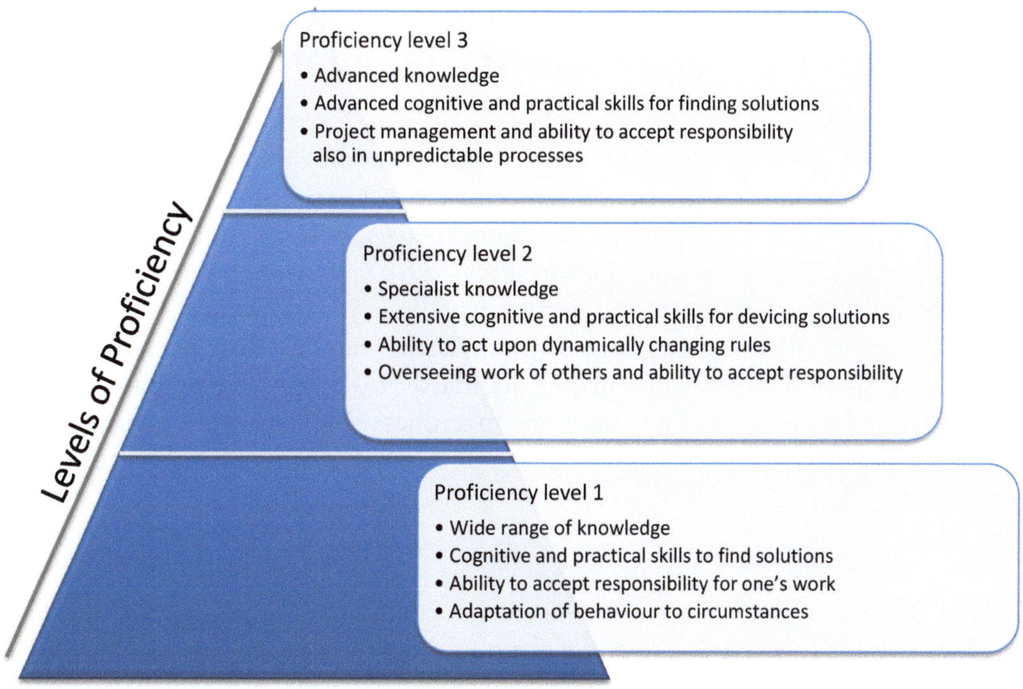

Figure 20: Levels of proficiency of persons creating information for use

10.4 Competencies of translators

The standard requires that translations of information for use, including checking and proofreading, must be handled by competent translators and language specialists.

The standard additionally recommends that translators should have the competencies of proficiency level 1 (see CHAPTER 10.3). They should be fluent in both the original and the target language. They should be native speakers of the target language and should be familiar with the type of product and any product-specific terminology.

11 The relation between IEC/IEEE 82079-1 and product or sector-specific standards

As a horizontal standard IEC/IEEE 82079-1 addresses all kinds of products. This is independent of whether the product is meant for industrial use (processing plant, industrial robot etc.) or for use by consumers (toaster, defibrillator for use by non-medical personnel) or even a service (see CHAPTER 2).

The standard provides basic and very comprehensive requirements on the principles for information quality, but without addressing product-specific details. This allows for interfaces with further standards to be applied parallel to IEC/IEEE 82079-1, such as EN ISO 20607 for machinery, ISO/IEC 26514 for software products, ISO 20417 for medical products and numerous others (see Figure 21).

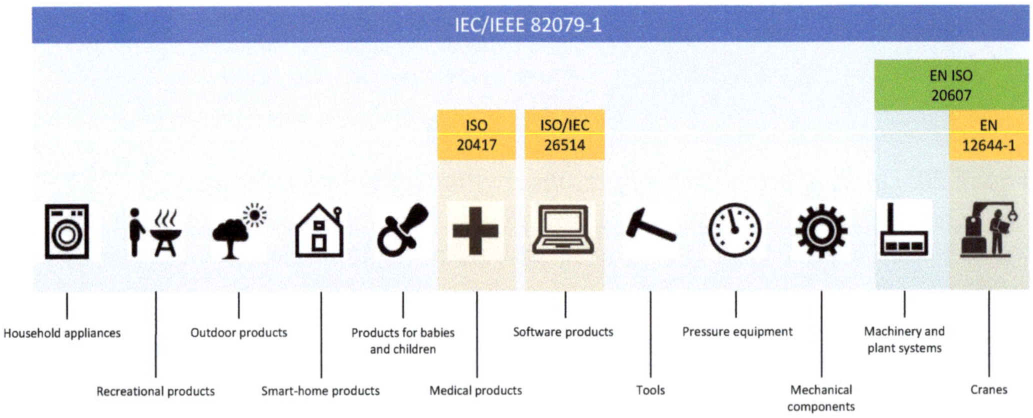

Figure 21: Interfaces between IEC/IEEE 82079-1 and other standards (excerpt)

Generally, product standardization committees should consider the basic principles and general requirements of applicable horizontal standards and include normative references to them in their product-specific standards (normative reference, see CHAPTER 2.5).

Product-specific standards should thus only standardize contents that are considered unique to the respective product group or a specific product and which must be added to the requirements of the horizontal standard, or which specify well-founded deviations.

However, in reality this principle is unfortunately rarely followed or observed to full extent. The main reason for this is the great number of horizontal standards, which are not all known to the individual product standardization committee.

Thus, if a product specific standard refers to IEC/IEEE 82079-1 informatively only or not at all regarding information for use, it can be concluded that the standard has not been considered sufficiently during preparation of the product-specific standard.

In such cases, therefore, possible overlap and even contradictions need to be traced down, to be able to fulfil the product-specific normative requirements. Overlaps and contradictions between the standards that need to be resolved, can only be determined by comparing the requirements.

In doing so, either a top-down or bottom-up principle may generally be followed.

Note:
Applying the top-down principle means proceeding from the abstract, superordinate level step by step to concrete, special, subordinate items.
Conversely, applying the bottom-up principle means beginning with concrete, special, subordinate items and proceeding step by step toward the abstract, general, superordinate level.
The top-down and bottom-up principles are also applied in fields such as hazard and risk analysis as well as in FMEA and general product design.
Following the top-down principle, the applicable requirements of superordinate standardization documents (here IEC/IEEE 82079-1) are collected first; then these are supplemented step by step with the more precise requirements from product and sector-specific standards.
Following the bottom-up principle, it is first determined which specific standards must be applied to the respective product and the requirements contained in them are collected. These are then supplemented with requirements from the superordinate standards such as IEC/IEEE 82079-1.
In practical application, both approaches are feasible and produce the desired results. However, note that the product-specific requirements must not simply be replaced by requirements from the basic standards.

Comparing the standards, it can be determined where individual documents include more detailed requirements. Applying the findings of such preliminary work allows ensuring that all relevant requirements for the information products to be prepared are observed. The remainder of this chapter illustrates this by compariing IEC/IEEE 82079-1 and EN ISO 20607 "Safety of machinery – Instruction handbook".

11.1 Product and sector-specific requirements take priority over general requirements

IEC/IEEE 82079-1, as a horizontal standard, contains more comprehensive requirements for the contents of information for use (see CHAPTER 7), requirements for the information management process (see CHAPTER 6), requirements for basic principles for preparation of information for use (see CHAPTER 5) as well as requirements for safety-related content (see CHAPTER 7.11).

By fulfilling these requirements of the standard, information for use will reach a high level, appropriate quality will be ensured and sustainable processes for their preparation likely be established.

Product and sector-specific standards often provide few generally applicable requirements. The contents of these specific standards are limited to the immediate applicability to a specific product or product group.

Connections to generally applicable horizontal standards are represented by means of normative or informative references (see CHAPTER 2.3). These cross-references are an important aid in applying the standards, because they allow identifying interfaces.

In many cases, product or sector-specific standards do not contain sufficient information for design of an all-embracing information development process. In such cases, IEC/IEEE 82079-1, as a horizontal standard, should be considered. Figure 22 schematically illustrates the approach to researching and applying the requirements from diverse standards.

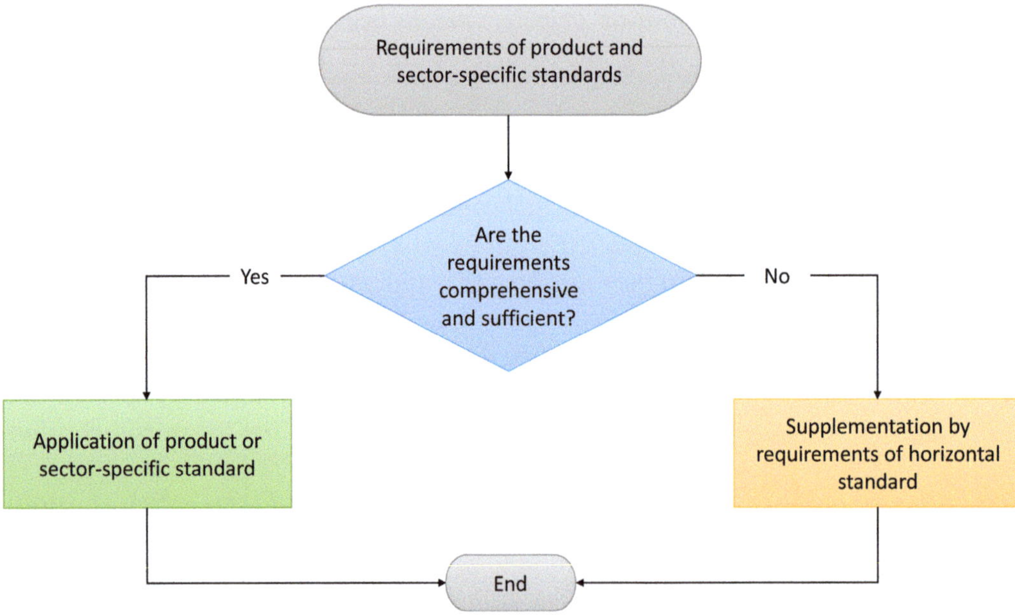

Figure 22: Researching and applying the requirements from diverse standards

Examples:
- EN ISO 20607 includes rudimentary requirements for preparation of information for use and limits itself to safety-related content of instruction handbooks for machinery. The horizontal standard should be applied additionally to benefit from the more comprehensive requirements for the information management process and the information quality principles.
- ISO 20417 on the other hand includes much more detailed requirements for information for use for medical products. In this case, the horizontal standard is a meaningful supplement to the requirements imposed upon medical products.

If product and sector-specific standards have been harmonised under national or international legislation (directives, regulations), it is imperative to primarily fulfil the requirements of these documents. If that is not sufficient to ensure careful design of information for use, the requirements of the horizontal standard IEC/IEEE 82079-1 should be applied additionally.

11.2 Sample presentation of the interfaces between EN ISO 20607 and IEC/IEEE 82079-1

EN ISO 20607 is a standard governing the preparation of information for use for machinery, and thus it is a product-specific standard. Already the scope of the standard says that general principles for concept and presentation of information for use can be found in IEC/IEEE 82079-1.

This is an indication regarding the way the standard should be used, clarifying that in general both standards should be applied (follow the approach presented in

Figure 22). General requirements are thus presented in IEC/IEEE 82079-1, whereas EN ISO 20607 includes only basic requirements for safety-related content.

11.2.1 Principles and general information

EN ISO 20607 requires that instruction handbooks for machinery contain information enabling users to operate the machinery safely and efficiently during the entire life cycle. The standard emphasizes a simplistic communication process of *Read - Think - Use*, considering the target audience and reasonably foreseeable misuse (see CHAPTER 7.8.1). The standard requires that information be presented so that it encourages correct and safe action following rather short pondering.

As target audiences EN ISO 20607 covers only a limited scope of persons. Definitely more detailed requirements for target audience orientation in information products are provided by IEC/IEEE 82079-1 (in subsection 6.2.2) as well as by ISO/IEC 26514.

EN ISO 20607 limits itself to a minimum of basic principles for preparation of content. Therefore, the principles presented in IEC/IEEE 82079-1 should be followed (see CHAPTER 5).

EN ISO 20607 describes in a little more detail how instruction handbooks may be published. According to it, information for use can be supplied

- in a single document or
- as a set of several individual documents

Assignment of contents to several individual documents may be useful, if

- different content is meant for differing target audiences or
- if a single document containing everything would be too complex and confusing
- if individual documents can be prepared for different phases of life or
- if different media are employed, as required by the users.

Thus the operating instructions for a user interface or the setting instructions for a special switching system may be provided in independent documents, while at the same time being a part of the entire instruction handbook.

11.2.2 Contents and structure of instruction handbooks

EN ISO 20607 provides a pre-fabricated structure with pre-defined contents (see *section 5*

The horizontal standard IEC/IEEE 82079-1 does not include a ready-made structuring proposal, but rules for structuring of contents (see CHAPTER 8).

Table 10 contains a comparison of contents and structure of IEC/IEEE 82079-1 and EN ISO 20607.

Table 10: Comparison of contents and structure of IEC/IEEE 82079-1 and EN ISO 20607.

IEC/IEEE 82079-1 **no ready-made proposal**	**EN ISO 20607** **Structuring proposal for the instruction handbook**	
Rules	**Chapter**	**Contents**
See CHAPTER 7 and CHAPTER 8. The structure of an information product must be adapted to diverse criteria, such as external influences (environment, illumination, application situations etc.), target audience(s), cognition, phase of life, and publication medium. Example: For digitally published information products, the structure is secondary, but the accessibility of the information by means of search functions is important. Structures are important on all levels: – information landscape – information products – information types (descriptive, procedural, referential) Proper structuring needs competent technical communicators (see CHAPTER 10).	Basic components of the instruction handbook	Title page Table of contents Introduction to and explanation of the handbook
	Safety	General safety information and general safety instructions
	Machine overview	Machine description Intended use Key machine specifications Control and display descriptions Layout and floor plan
	Transport, handling and storage	Transporting, handling and storing the machine and/or the components
	Assembly, installation and commissioning	Assembly/integration of the machine Positioning of the machine Mechanical, pneumatic, hydraulic and electrical installation Check and test of safety systems Installation check Commissioning
	Original equipment manufacturer settings	Mechanical settings and synchronization Safety-related (setting) parameters Pneumatic, hydraulic, electrical and vacuum settings Other settings
	Operation	Operating modes Machine start-up and shut-down Sequence or chronology of operations Other operating instructions, (if applicable)

IEC/IEEE 82079-1 no ready-made proposal	**EN ISO 20607 Structuring proposal for the instruction handbook**	
Rules	**Chapter**	**Contents**
	Product or capacity changeover	General product or capacity changeover information
		Product-specific set-up information
	Inspection, testing and maintenance	Maintenance instructions
	Cleaning and sanitizing	Cleaning and/or disinfection of the machinery
	Fault finding/ troubleshooting and repair	General fault finding/troubleshooting and repair information
		Troubleshooting chart (electrical installation)
		Troubleshooting of electrical sensors, vacuum systems, pneumatic systems and hydraulic systems
	Dismantling, disabling and scrapping	Dismantling, disabling and scrapping instructions
	Documents and drawings	Further documents, drawings, and spare parts list
	Index, glossary, annexes	

11.2.3 IT security in instruction handbooks

EN ISO 20607 requires that the information for use contains information on IT security systems operating in conjunction with the product. All product internal security systems and the IT interfaces must be considered. The information provided should guarantee and maintain the security of the product against influences such as hacking. All conditions enabling the user to maintain control of the security inside and in connection with IT systems must be described.

Information products must be so designed that, based on previous risk assessment, IT vulnerabilities relevant for the customer/user can be detected.

Regarding this topic, EN ISO 20607 points to the technical report ISO/TR 22100-4. It presents the requirements of ISO 12100 on IT security in greater detail. The requirements presented also pertain to preparation of information for use and thus substantially influence the preparation process.

When information for use is prepared according to EN ISO 20607, the following contents concerning IT safety should be included,therefore:

- Information on user rights and restriction of these (assignment of roles)
- Instruction and information on preparation of individual user accounts and the related account management
- Instructions on risk reduction measures, such as updating of passwords, firewalls and anti-virus software etc.

- Information for use on authentication and access control mechanisms, e.g. card readers, conventional, electronic or digital locks)
- Instructions on installation of updates
- Measures guaranteeing IT security, e.g. in case of access for remote maintenance etc.
- Information on regular system check-up and detection of failing system components or unavailable system services

These requirements largely cover the requirements of IEC/IEEE 82079-1 (see CHAPTER 7.14).

In applying either standard, it is essential to consider safety aspects from the risk assessment when preparing the information for use.

11.2.4 Forms of publication approved by the standards

EN ISO 20607 and IEC/IEEE 82079-1 agree concerning the delivery of information products by means of different media and through different channels (see CHAPTER 9).

In agreement with the applicable legal requirements, EN ISO 20607 for machinery approves of the publication of instructions by means of different media and through diverse channels:
- publication of instructions in printed form
- delivery on an electronic storage medium made available with the machine
- delivery by means of access to external servers, websites or other storage (e.g. cloud-based storage etc.)
- delivery on internal servers or storage media
- delivery as audio or video instruction running on control terminals, websites, as audio files or for download from the Internet

If special requirements are set by public regulations (laws, ordinances), these must be observed. In absence of legal regulations, a contract agreement on the form of publication should be made with the user (customer).

Note:
EN ISO 20607 mentions the same requirement concerning the language versions of instruction handbooks. If there are legal requirements, these must be fulfilled. Only in absence of applicable legal regulations, the language version should be agreed upon with the user (customer).

In choosing the form of publication, the intended use, the target audience(s) and the operating environment should be considered. If delivery in digital form cannot be guaranteed, a printed version must be at hand for use (see CHAPTER 9.2 and CHAPTER 9.4).

11.3 EN ISO 20607 and IEC/IEEE 82079-1 compared side by side

Since the scope of application is limited to machinery in EN ISO 20607, IEC/IEEE 82079-1 includes more detailed requirements and suggestions for the preparation of information for use.

The overlap of both standards mainly concerns formal requirements and some content matter. The comparison clearly highlights where the standards mutually supplement each other, and where requirements agree (see Table 11).

Table 11: Requirements of EN ISO 20607 and IEC/IEEE 82079-1 compared side by side

Topic	EN ISO 20607	IEC/IEEE 82079-1	Comparison
Identification of the information product	Date of issue and version identification, identification number, serial number(s) and period of application	Identification number and date of issue of the information product	agreeing
Identification of the product	Machine designation (model and/or type)	Name, information about configuration	agreeing; IEC/IEEE 82079-1 provides more detailed requirements
Identification of supplier	Name of the manufacturer and contact details (address, phone numbers, email address, website)	Name, address, telephone numbers, email addresses, hyper-links	agreeing
Identification of supplier	–	Contact details of certain providers and of customer service	agreeing; IEC/IEEE 82079-1 provides more detailed requirements
Format/medium	Broad range (paper and electronic/digital media)	Broad range (paper and electronic/digital media)	compatible; but more detailed in IEC/IEEE 82079-1
Structure (subdivision of information in instructions)	Template for contents with "sections", concrete structuring proposal	Rules and criteria for structuring; listing of contents, no structuring proposal	different approaches that complement another
Safety notes and warning messages	Few requirements, but outline for structuring of warning messages	Consistent concept in agreement with ANSI Z535	compatible; IEC/IEEE 82079-1 provides more detailed requirements
Failures	Suggestion for structure of a troubleshooting table	Basic requirements	supplementary; EN ISO 20607 provides a sample for the contents

Topic	EN ISO 20607	IEC/IEEE 82079-1	Comparison
Text	Wording rules, e.g. imperative for instructions (steps), avoiding negative words, short sentences, understandable words etc.	Simple wording, consistent terminology, unambiguous function of sections of information	supplementary
Text font size	Table from IEC/IEEE 82079-1 has been taken over	Table with detailed recommendations	agreeing
Information security	Cross-reference to ISO/TR 22100-4	Requires including corresponding instructions and considering common methods	supplementary, if ISO/TR 22100-4 is applied

11.4 Applying several standards at a time

If several standards are applied at the same time, overlapping and supplementary requirements must be noted (see CHAPTER 2).

When preparing information for use, the product-specific requirements should be observed first. If these are not sufficient for the use case at hand, further requirements, e.g. from horizontal standards serve closing the gaps left by the more specific standards (also see CHAPTER 11.1 and Figure 22).

To achieve a structured overview, the requirements should be subdivided into the following three ranges:

1. requirements for the design and preparation of information for use
2. requirements for the content of information for use
3. supplementary requirements concerning the creating process

This subdivision should result in a neat summary, facilitating orderly clarification of the numerous overlapping requirements. Only such well-organized processing of requirements from several standards applied in parallel will result in as high as possible a degree of fulfilment of common and supplementary requirements.

In technical editing departments, IEC/IEEE 82079-1 will always be the most important standard, despite numerous product and sector-specific standards. This holds true, even though there are numerous overlapping normative requirements.

12 Appendix

12.1 Requirements for the content of information for use set forth in the Machinery Directive (2006/42/EC)

Note:
Information for use is referred to as "instructions" in the Machinery Directive.

Section of MD	Requirement	Contained in instructions	
		yes (chp.)	no/n. a.
Annex I, chapter 1.7.4.2: Contents of the instructions			
a	the business name and full address of the manufacturer and of his authorised representative		
b	the designation of the machinery as marked on the machinery itself, except for the serial number (see section 1.7.3)		
c	the EC declaration of conformity, or a document setting out the contents of the EC declaration of conformity, showing the particulars of the machinery, not necessarily including the serial number and the signature		
d	a general description of the machinery (see CHAPTER 7.8.1 of the present guide)		
e	the drawings, diagrams, descriptions and explanations necessary for the use, maintenance and repair of the machinery and for checking its correct functioning		
f	a description of the workstation(s) likely to be occupied by operators		
g	a description of the intended use of the machinery		
h	warnings concerning ways in which the machinery must not be used that experience has shown might occur		
i	assembly, installation and connection instructions, including drawings, diagrams and the means of attachment and the designation of the chassis or installation on which the machinery is to be mounted		
j	instructions relating to installation and assembly for reducing noise or vibration		
k	instructions for the putting into service and use of the machinery and, if necessary, instructions for the training of operators		
l	information about the residual risks that remain despite the inherent safe design measures, safeguarding and complementary protective measures adopted		
m	instructions on the protective measures to be taken by the user, including, where appropriate, the personal protective equipment to be provided		
n	the essential characteristics of tools which may be fitted to the machinery		

Section of MD	Requirement	Contained in instructions	
		yes (chp.)	no/n.a.
o	the conditions in which the machinery meets the requirement of stability during use, transportation, assembly, dismantling when out of service, testing or foreseeable breakdowns		
p	instructions with a view to ensuring that transport, handling and storage operations can be made safely, giving the mass of the machinery and of its various parts where these are regularly to be transported separately		
q	the operating method to be followed in the event of accident or breakdown; if a blockage is likely to occur, the operating method to be followed so as to enable the equipment to be safely unblocked		
r	the description of the adjustment and maintenance operations that should be carried out by the user and the preventive maintenance measures that should be observed		
s	instructions designed to enable adjustment and maintenance to be carried out safely, including the protective measures that should be taken during these operations		
t	the specifications of the spare parts to be used, when these affect the health and safety of operators		
u	the following information on airborne noise emissions		
v	where machinery is likely to emit non-ionising radiation which may cause harm to persons, in particular persons with active or non-active implantable medical devices, information concerning the radiation emitted for the operator and exposed persons		

12.2 Limits of machinery

To be able to describe the limits of the machinery, the person preparing the information for use needs to find the following information in the risk assessment. This listing is an example and we do not claim it is complete. Depending on the product, additional information may be needed.

Use limits
- operating modes
- range of application (indoors, outdoors, industry, private settings etc.)
- level of training of operators
- oral instruction of persons not operating the machine (e.g. visitors)

Space limits
- space required
- interfaces

Time limits
- service life
- maintenance intervals

Other limits
- properties of the material(s) to be processed
- Cleaning

12.3 Phases of life of the machinery

Examples for the phases of life of machinery:
- planning
- design
- assembly
- commissioning
  - ▹ operation (use)
  - ▹ normal operation
  - ▹ setting
  - ▹ maintenance/service
  - ▹ trouble shooting
- de-commissioning
  - ▹ transient (e. g. for remedy of a failure)
  - ▹ long-term (e. g. to perform planned maintenance)
  - ▹ final de-commissioning
- disposal

12.4 Example for a table of contents

BHB Example

Contents

BHB Example

BHB Example

Figure 23: Example of a table of contents in the information for use of an industrial furnace

12.5 Examples of methods for determination of signal words

12.5.1 Example based on the risk graph by Reudenbach

					Probability of occurrence	Low / hardly possible			Medium / quite likely			Hogh / highly probable		
					W	W3 = 1			W4 = 3			W5 = 5		
					Avoidance	Possible	Conditionally possible	Impossible	Possible	Conditionally possible	Impossible	Possible	Conditionally possible	Impossible
S	Schwere	W	Fre-quency	Duration	W	W6 = 1	W7 = 2	W8 = 3	W6 = 1	W7 = 2	W8 = 3	W6 = 1	W7 = 2	W8 = 3
S1 = 1	No consequences	W1 = 1	seldom	short		3	4	5	5	6	7	7	8	9
		W2 = 2	often	long		4	5	6	6	7	8	8	9	10
S2 = 3	Minor consequences	W1 = 1	seldom	short		9	12	15	15	18	21	21	24	27
		W2 = 2	often	long		12	15	18	18	21	24	24	27	30
S3 = 6	Moderately severe consequences	W1 = 1	seldom	short		18	24	30	30	36	42	42	48	54
		W2 = 2	often	long		24	30	36	36	42	48	48	54	60
S4 = 8	Severe consequences, permanent damage possible	W1 = 1	seldom	short		24	32	40	40	48	56	56	64	72
		W2 = 2	often	long		32	40	48	48	56	64	64	72	80
S5 = 10	Fatal consequences	W1 = 1	seldom	short		30	40	50	50	60	70	70	80	90
		W2 = 2	often	long		40	50	60	60	70	80	80	90	100

R	Risk index
0 – 24	Low risk
25 – 42	Medium risk
42 – 100	High risk

NOTICE
CAUTION
WARNING
DANGER

Computation:	**R = S x W** **R** = **S**(1,2,3,4,5) x ((**W**(1,2) + **W**(3,4,5) + **W**(6,7,8))

R:	Risk index
S (1, 2, 3, 4, 5):	Severity weighting
W (1, 2):	Frequency and duration weighting
W (3, 4, 5):	Probability of occurrence weighting
W (6, 7, 8):	Possibility of avoidance or limiting of consequences weighting

Figure 24: Example for determination of signal words based on the risk graph by Reudenbach

12.5.2 Example based on the risk graph from EN 60601

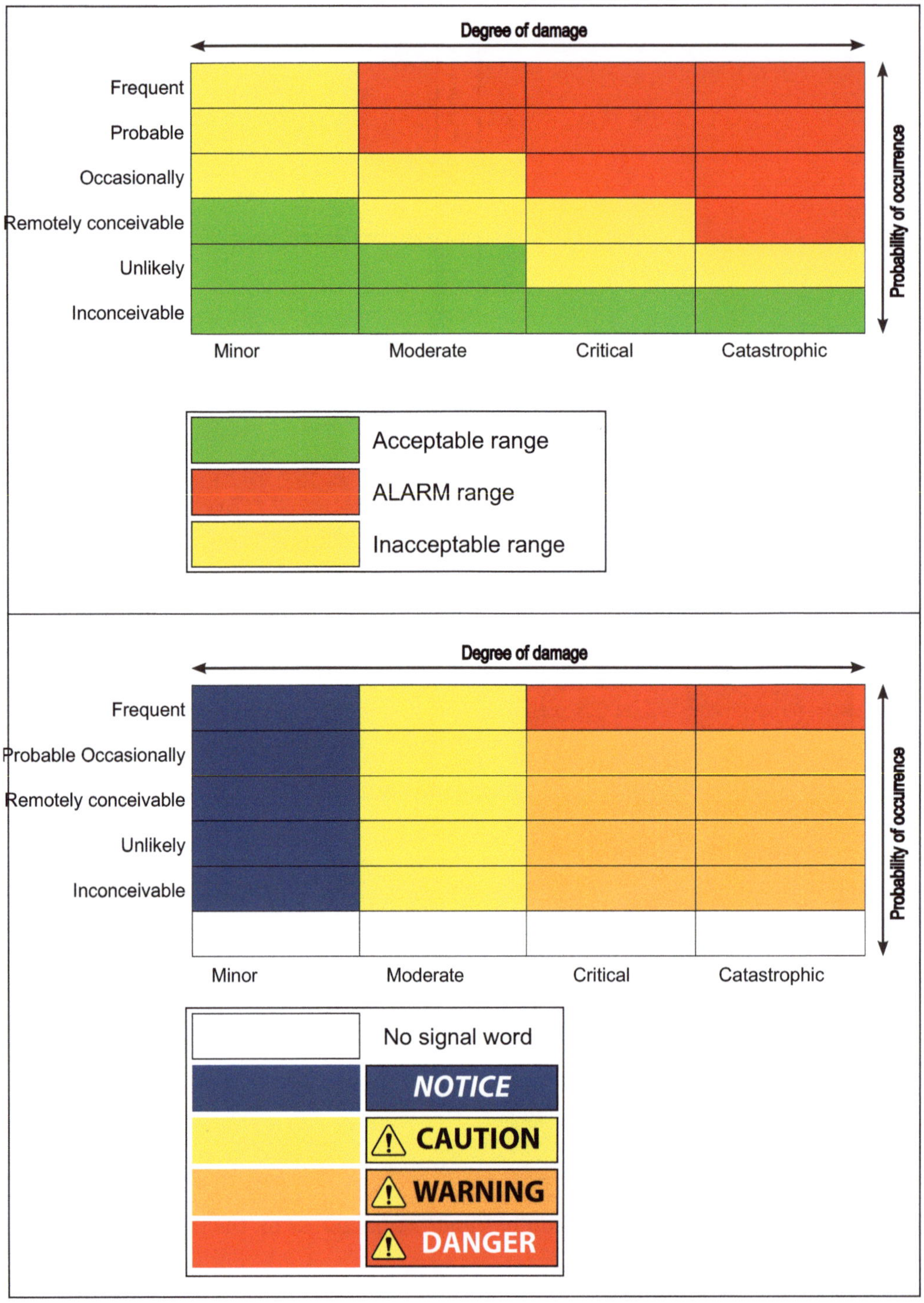

Figure 25: Example for determination of signal words based on the risk graph from EN 60601

13 Literature

ANSI Z535.4-2011 (R2017) Product Safety Signs and Labels

ANSI Z535.6-2011 (R2017) Product Safety Information in Product Manuals, Instructions and Other Collateral Materials

DIN 820-2:2018-09 Standardization - Part 2: Presentation of documents (ISO/IEC Directives) – >Part 2:2016, modified

EN 60601-1:2013-12 Medical electrical equipment - Part 1: General requirements for basic safety and essential performance

EN 61355-1:2009-03 Classification and designation of documents for plants, systems and equipment - Part 1: Rules and classification tables (IEC 61355-1:2008)

EN 82079-1:2013-06 Preparation of instructions for use - Structuring, content and presentation - Part 1: General principles and detailed requirements (IEC 82079-1:2012)

EN ISO 12100:2011-03 Safety of machinery - General principles for design - Risk assessment and risk reduction (ISO 12100:2010)

EN 12644-1 Cranes – Information for use and testing – Part 1: Instructions

EN ISO 20607:2019 Safety of machinery – Instruction handbook – General drafting principles

IEC 60073:2002 Basic and safety principles for man-machine interface, marking and identification – Coding principles for indicators and actuator

IEC 60335 Household and similar electrical appliances (standard series)

IEC 60417:2002 Graphical symbols for use on equipment (database)

IEC 60617:2012 Graphical symbols for diagrams (database)

IEC 60825-1:2014 Safety of laser products – Part 1: Equipment classification and requirements

IEC 61082-1:2014 Preparation of documents used in electro-technology – Part 1: Rules

IEC 61506:1997 Industrial-process measurement and control – Documentation of application software

IEC 62023:2011 Structuring of technical information and documentation

IEC 62744:2014 Representation of states of objects by graphical symbols

IEC 81346-1:2009 Industrial systems, installations and equipment and industrial products – Structuring principles and reference designations – Part 1: Basic rules

IEC/IEEE 82079-1 Preparation of information for use (instructions for use) of products – Part 1: Principles and general requirements

ISO 639-2:1998 Codes for the representation of names of languages – Part 2: Alpha-3 code

ISO 704:2009 Terminology work – Principles and methods

ISO 3864-1:2011 Graphical symbols – Safety colours and safety signs – Part 1: Design principles for safety signs and safety markings

ISO 3864-2:2016 Graphical symbols – Safety colours and safety signs – Part 2: Design principles for product safety labels

ISO 5807:1985 Information processing – Documentation symbols and conventions for data, program and system flowcharts, program network charts and system resources charts
ISO 5963:1985 Documentation – Methods for examining documents, determining their subjects, and selecting indexing terms
ISO 7000 Graphical symbols for use on equipment
ISO 7010 Graphical symbols – Safety colours and safety signs – Registered safety signs
ISO 7731:2003 Ergonomics – Danger signals for public and work areas – Auditory danger signals
ISO 9241-300:2008 Ergonomics of human-system interaction – Part 300: Introduction to electronic visual display requirements
ISO 11429:1996 Ergonomics – System of auditory and visual danger and information signals
ISO 14617-1:2005 Graphical symbols for diagrams – Part 1: General information and indexes
ISO 15006:2011 Road vehicles – Ergonomic aspects of transport information and control systems – Specifications for in-vehicle auditory presentation
ISO 15519-1:2010 Specification for diagrams for process industry – Part 1: General rules
ISO/TR 16352:2005 Road vehicles – Ergonomic aspects of in-vehicle presentation for transport information and control systems – Warning systems
ISO 17100 Translation Services – Requirements for translation services
ISO/DIS 20417 Medical devices – Information to be provided by the manufacturer
ISO/TR 22100-4:2018 Safety of machinery – Relationship with ISO 12100 – Part 4: Guidance to machinery manufacturers for consideration of related IT-security (cyber security)
ISO 80000-11:2019 Quantities and units – Part 11: Characteristic numbers

ISO/IEC/IEEE 23026:2015 Systems and software engineering – Engineering and management of websites for systems, software, and services information
ISO/IEC 26514:2008 Systems and software engineering – Requirements for designers and developers of user documentation
ISO/IEC 27002 Information technology – Security techniques – Code of practice for information security controls
ISO/IEC 40500:2012 Information technology – W3C Web Content Accessibility Guidelines (WCAG) 2.0
ISO/IEC Directives Part 2: 2018 Principles and rules for the structure and drafting of ISO and IEC documents

tekom Richtlinie: Bereitstellung von Nutzungsinformationen in elektronischer Form – eDok (Delivery of information for use in electronic form), 2. Auflage, tcworld Verlag
VDI 4500 Part 6 Technical documentation - Documentation process - Publishing

14 Table of figures and tables

15 Authors

Martin Tillmann

Dipl.-Ing. Martin Tillmann studied mechanical engineering from 1985 to 1990 with the focus on energy and processing technology at Essen University.

Upon several professional career stations, for instance, as scientific assistant for power station research, technical operations manager of a malt factory and manager of the subsidiary of a technical documentation service provider, he founded Engineering & Dokumentation Tillmann in 2002, which became part of ED-Technik GmbH in 2019. This engineering service provider supports manufacturers and operating companies in the areas of CE Marking of products and preparation of technical documentation.

Martin Tillmann is a trained technical author, certified CE coordinator and expert for CE conformity assessment and technical documentation certified to EN ISO/IEC 17024.

Michael Fritz

Michael Fritz holds a doctorate in media sciences and a master's degree in business administration with a focus on marketing. He has been employed with various associations, for instance as the manager of the Institut für lebenslanges Lernen (institute for lifelong learning).

Since 1996 he is the managing director of the Gesellschaft für Technische Kommunikation – tekom Deutschland e.V. and since 2013 also the managing director of Europe's largest association of technical communicators – tekom Europe. Additionally, he is the managing director of tcworld GmbH, a service provider for tekom/tekom Europe.

Claudia Klumpp

Dr. jur. Claudia Klumpp is a legal professional and holds a degree in business administration (VWA).

After years of work as a lawyer and legal consultant, she is working for tekom since 2013 and – together with tekom´s Advisory Board for Legislation and Standards – is responsible for the involvement in standardization of tekom Deutschland and tekom Europe.

Since 2016 Claudia Klumpp as convenor chairs the international working group (IEC TC 3/ISO TC 10 JWG 16), who has revised IEC/IEEE 82079-1. Additionally, she chairs the national mirror committee for IEC/IEEE 82079-1 as the convenor.

Since 2016 she works as a lecturer for "Rechtliche Grundlagen und Normen" (legal basics and standardization) in the extra-occupational degree course "Technical Documentation" at Karlsruhe University.

Martin Rieder

Martin Rieder is the managing director of the engineering service provider CAVEO Safety Mangagement & Documentation, an industrial engineer, technical communicator and expert for technical safety and documentation. He is offering consultancy for product safety, conformity and liability as well as for technical documentation to customers worldwide.

He is a member of the board of tekom Europe and tekom Austria; he also plays a mayor role in national and international standardization. As an expert of tekom´s Advisory Board for Legislation and Standards he is dealing with topics such as product conformity and the revision of the Machinery Directive.

He is pleased to share his knowledge as a lecturer of the master course for Technical Documentation (FH Joanneum), as a trainer, speaker and author of technical books.

Roland Schmeling

Roland Schmeling, Dipl.-Phys., is the managing director of the consultancy SCHMELING + CONSULTANTS in Heidelberg.

Since 1999 he is providing consultancy services in the field of technical communication and documentation with a focus on standardization, information concepts, conformity and change. He studied physics and analytical philosophy. Since 2004 he teaches quality assurance at Furtwangen University. As an auditor of TÜV SÜD he inspects and audits documentations and editorial processes to the DocCert standard.

For tekom Roland Schmeling participates in diverse working groups, committees and preparation of publications. As a voluntary member of the Advisory Board for Legislation and Standards he participates in the work on IEC/IEEE 82079-1 both on the national and international level.

Stephan Schneider

Stephan Schneider, Dipl.-Ing. (FH), studied electrical energy technologies from 1984 to 1989 at Konstanz University of Applied Sciences (FH).

Since 1995 he is working for SEW-EURODRIVE and is the manager of the technical editing department since 2002. His responsibilities include the strategic and conceptional development of technical communication.

Before, he was employed as a development engineer in a medium-sized mechanical engineering company.

IEC/IEEE 82079-1

Edition 2.0 2019-05

INTERNATIONAL STANDARD
NORME INTERNATIONALE

colour inside

HORIZONTAL STANDARD
NORME HORIZONTALE

Preparation of information for use (instructions for use) of products –
Part 1: Principles and general requirements

Élaboration des informations d'utilisation (instructions d'utilisation) des produits –
Partie 1: Principes et exigences générales

IEC/IEEE 82079-1:2019-05(en-fr)

IEC Central Office
3, rue de Varembé
CH-1211 Geneva 20
Switzerland
Tel.: +41 22 919 02 11
info@iec.ch
www.iec.ch

Institute of Electrical and Electronics Engineers, Inc.
3 Park Avenue
New York, NY 10016-5997
United States of America
stds.info@ieee.org
www.ieee.org

About the IEC
The International Electrotechnical Commission (IEC) is the leading global organization that prepares and publishes International Standards for all electrical, electronic and related technologies.

About IEC publications
The technical content of IEC publications is kept under constant review by the IEC. Please make sure that you have the latest edition, a corrigendum or an amendment might have been published.

IEC publications search - webstore.iec.ch/advsearchform
The advanced search enables to find IEC publications by a variety of criteria (reference number, text, technical committee,...). It also gives information on projects, replaced and withdrawn publications.

IEC Just Published - webstore.iec.ch/justpublished
Stay up to date on all new IEC publications. Just Published details all new publications released. Available online and once a month by email.

IEC Customer Service Centre - webstore.iec.ch/csc
If you wish to give us your feedback on this publication or need further assistance, please contact the Customer Service Centre: sales@iec.ch.

Electropedia - www.electropedia.org
The world's leading online dictionary on electrotechnology, containing more than 22 000 terminological entries in English and French, with equivalent terms in 16 additional languages. Also known as the International Electrotechnical Vocabulary (IEV) online.

IEC Glossary - std.iec.ch/glossary
67 000 electrotechnical terminology entries in English and French extracted from the Terms and Definitions clause of IEC publications issued since 2002. Some entries have been collected from earlier publications of IEC TC 37, 77, 86 and CISPR.

A propos de l'IEC
La Commission Electrotechnique Internationale (IEC) est la première organisation mondiale qui élabore et publie des Normes internationales pour tout ce qui a trait à l'électricité, à l'électronique et aux technologies apparentées.

A propos des publications IEC
Le contenu technique des publications IEC est constamment revu. Veuillez vous assurer que vous possédez l'édition la plus récente, un corrigendum ou amendement peut avoir été publié.

Recherche de publications IEC - webstore.iec.ch/advsearchform
La recherche avancée permet de trouver des publications IEC en utilisant différents critères (numéro de référence, texte, comité d'études,...). Elle donne aussi des informations sur les projets et les publications remplacées ou retirées.

IEC Just Published - webstore.iec.ch/justpublished
Restez informé sur les nouvelles publications IEC. Just Published détaille les nouvelles publications parues. Disponible en ligne et une fois par mois par email.

Service Clients - webstore.iec.ch/csc
Si vous désirez nous donner des commentaires sur cette publication ou si vous avez des questions contactez-nous: sales@iec.ch.

Electropedia - www.electropedia.org
Le premier dictionnaire d'électrotechnologie en ligne au monde, avec plus de 22 000 articles terminologiques en anglais et en français, ainsi que les termes équivalents dans 16 langues additionnelles. Egalement appelé Vocabulaire Electrotechnique International (IEV) en ligne.

Glossaire IEC - std.iec.ch/glossary
67 000 entrées terminologiques électrotechniques, en anglais et en français, extraites des articles Termes et Définitions des publications IEC parues depuis 2002. Plus certaines entrées antérieures extraites des publications des CE 37, 77, 86 et CISPR de l'IEC.

IEC/IEEE 82079-1

Edition 2.0 2019-05

INTERNATIONAL STANDARD
NORME INTERNATIONALE

colour inside

HORIZONTAL STANDARD
NORME HORIZONTALE

Preparation of information for use (instructions for use) of products – Part 1: Principles and general requirements

Élaboration des informations d'utilisation (instructions d'utilisation) des produits – Partie 1: Principes et exigences générales

INTERNATIONAL
ELECTROTECHNICAL
COMMISSION

COMMISSION
ELECTROTECHNIQUE
INTERNATIONALE

ICS 01.110; 29.020

ISBN 978-2-8322-6835-3

Warning! Make sure that you obtained this publication from an authorized distributor.
Attention! Veuillez vous assurer que vous avez obtenu cette publication via un distributeur agréé.

CONTENTS

INTERNATIONAL ELECTROTECHNICAL COMMISSION

____

PREPARATION OF INFORMATION FOR USE (INSTRUCTIONS FOR USE) OF PRODUCTS –

Part 1: Principles and general requirements

FOREWORD

1) The International Electrotechnical Commission (IEC) is a worldwide organization for standardization comprising all national electrotechnical committees (IEC National Committees). The object of IEC is to promote international co-operation on all questions concerning standardization in the electrical and electronic fields. To this end and in addition to other activities, IEC publishes International Standards, Technical Specifications, Technical Reports, Publicly Available Specifications (PAS) and Guides (hereafter referred to as "IEC Publication(s)"). Their preparation is entrusted to technical committees; any IEC National Committee interested in the subject dealt with may participate in this preparatory work. International, governmental and non-governmental organizations liaising with the IEC also participate in this preparation.

   IEEE Standards documents are developed within IEEE Societies and Standards Coordinating Committees of the IEEE Standards Association (IEEE-SA) Standards Board. IEEE develops its standards through a consensus development process, which brings together volunteers representing varied viewpoints and interests to achieve the final product. Volunteers are not necessarily members of IEEE and serve without compensation. While IEEE administers the process and establishes rules to promote fairness in the consensus development process, IEEE does not independently evaluate, test, or verify the accuracy of any of the information contained in its standards. Use of IEEE Standards documents is wholly voluntary. IEEE documents are made available for use subject to important notices and legal disclaimers (see http://standards.ieee.org/IPR/disclaimers.html for more information).

   IEC collaborates closely with IEEE in accordance with conditions determined by agreement between the two organizations.

2) The formal decisions of IEC on technical matters express, as nearly as possible, an international consensus of opinion on the relevant subjects since each technical committee has representation from all interested IEC National Committees. The formal decisions of IEEE on technical matters, once consensus within IEEE Societies and Standards Coordinating Committees has been reached, is determined by a balanced ballot of materially interested parties who indicate interest in reviewing the proposed standard. Final approval of the IEEE standards document is given by the IEEE Standards Association (IEEE-SA) Standards Board.

3) IEC/IEEE Publications have the form of recommendations for international use and are accepted by IEC National Committees/IEEE Societies in that sense. While all reasonable efforts are made to ensure that the technical content of IEC/IEEE Publications is accurate, IEC or IEEE cannot be held responsible for the way in which they are used or for any misinterpretation by any end user.

4) In order to promote international uniformity, IEC National Committees undertake to apply IEC Publications (including IEC/IEEE Publications) transparently to the maximum extent possible in their national and regional publications. Any divergence between any IEC/IEEE Publication and the corresponding national or regional publication shall be clearly indicated in the latter.

5) IEC and IEEE do not provide any attestation of conformity. Independent certification bodies provide conformity assessment services and, in some areas, access to IEC marks of conformity. IEC and IEEE are not responsible for any services carried out by independent certification bodies.

6) All users should ensure that they have the latest edition of this publication.

7) No liability shall attach to IEC or IEEE or their directors, employees, servants or agents including individual experts and members of technical committees and IEC National Committees, or volunteers of IEEE Societies and the Standards Coordinating Committees of the IEEE Standards Association (IEEE-SA) Standards Board, for any personal injury, property damage or other damage of any nature whatsoever, whether direct or indirect, or for costs (including legal fees) and expenses arising out of the publication, use of, or reliance upon, this IEC/IEEE Publication or any other IEC or IEEE Publications.

8) Attention is drawn to the normative references cited in this publication. Use of the referenced publications is indispensable for the correct application of this publication.

9) Attention is drawn to the possibility that implementation of this IEC/IEEE Publication may require use of material covered by patent rights. By publication of this standard, no position is taken with respect to the existence or validity of any patent rights in connection therewith. IEC or IEEE shall not be held responsible for identifying Essential Patent Claims for which a license may be required, for conducting inquiries into the legal validity or scope of Patent Claims or determining whether any licensing terms or conditions provided in connection with submission of a Letter of Assurance, if any, or in any licensing agreements are reasonable or non-discriminatory. Users of this standard are expressly advised that determination of the validity of any patent rights, and the risk of infringement of such rights, is entirely their own responsibility.

International Standard IEC/IEEE 82079-1 has been prepared by IEC technical committee 3: Information structures and elements, identification and marking principles, documentation and graphical symbols, in cooperation with the Computer Society, Systems and Software Engineering Standards Committee of the IEEE, under the IEC/IEEE Dual Logo Agreement and in cooperation with subcommittee 1: Basic conventions of ISO technical committee 10: Technical product documentation.

It is published as an IEC/ISO/IEEE triple logo standard.

It has the status of a horizontal standard in accordance with IEC Guide 108 [59].

This second edition cancels and replaces the first edition published in 2012. This edition constitutes a technical revision.

This edition includes the following significant technical changes with respect to the previous edition:

a) The structure of this document has been rearranged in order to facilitate application of the standard and to make it easier to find information. Where possible, the language has been simplified.
b) Information for use is introduced as a generic term. Instructions for use is a synonym for information for use. Step-by-step instructions is used as a subset of information for use.
c) Clause 5 (principles) is revised and focuses on the purpose of information for use, the quality of information and the process for management of information.
d) The process for preparation of information for use is integrated in the normative part and addressed comprehensively.
e) Empirical methods for the evaluation of information for use are described in the normative part.
f) The professional competencies needed for the preparation of information for use are addressed more comprehensively.
g) Some aspects have been added to general requirements for information for use for complex systems of systems.
h) Consideration is given to instructions for self-assembly products.
i) An informative annex providing guidance on the fulfilment of specified requirements is introduced.

The text of this International Standard is based on the following IEC documents:

FDIS	Report on voting
3/1390/FDIS	3/1401/RVD

Full information on the voting for the approval of this International Standard can be found in the report on voting indicated in the above table.

International Standards are drafted in accordance with the rules given in the ISO/IEC Directives, Part 2.

A list of all parts of the 82079 International Standard, published under the general title *Preparation of information for use (instructions for use) of products*, can be found on the IEC website.

The committee has decided that the contents of this document will remain unchanged until the stability date indicated on the IEC website under "http://webstore.iec.ch" in the data related to the specific document. At this date, the document will be

- reconfirmed,
- withdrawn,
- replaced by a revised edition, or
- amended.

IMPORTANT – The 'colour inside' logo on the cover page of this publication indicates that it contains colours which are considered to be useful for the correct understanding of its contents. Users should therefore print this document using a colour printer.

INTRODUCTION

Information for use is a part of any type of product it supports. A product can be a system, a service, goods, software, information, or a combination thereof. People depend on the information provided to use products safely, effectively, and efficiently, unless they receive training from a human instructor or unless the functions are entirely intuitive. Confusing product information and inadequate instructions are major sources of frustration for consumers and skilled workers. Defective information can pose a risk of harm or loss, leading to prosecution or liability claims against the supplier or brand owner.

Information for use consists of three information types: conceptual information that the target audience needs to understand, instructional information to be followed or considered, and reference information to be consulted when needed. The information for use can include various information products that are selected, presented, and delivered on different media to meet the needs of different target audiences (Figure 1).

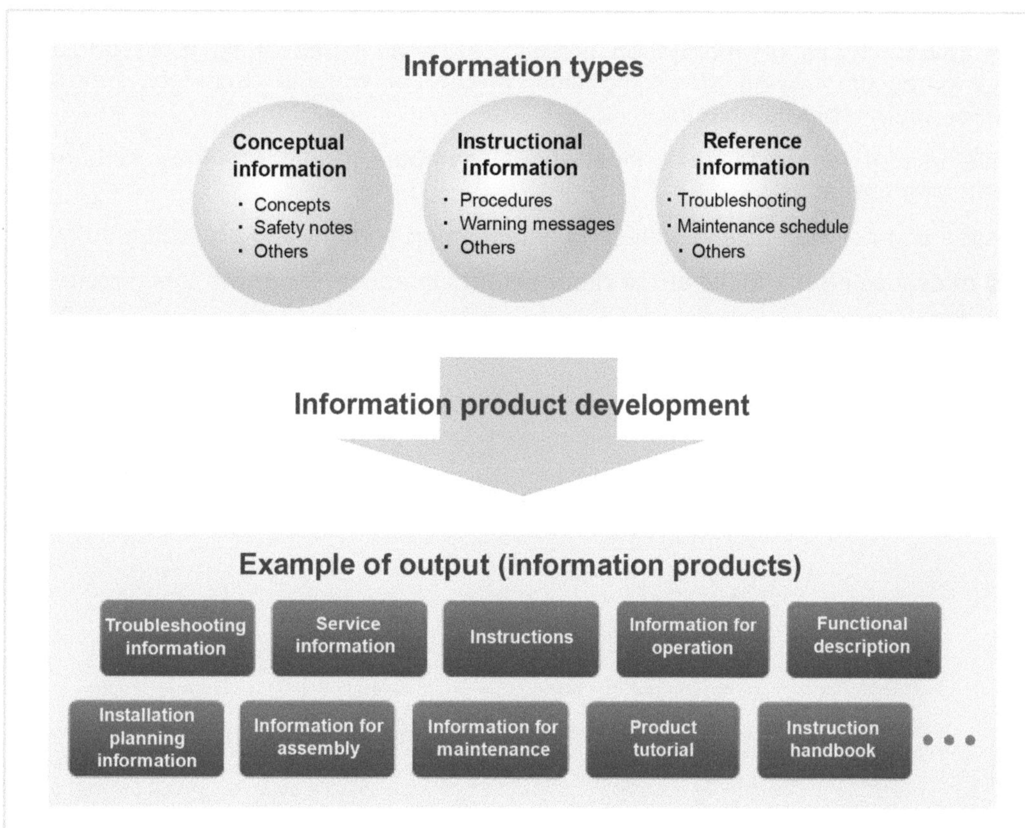

IEC

Figure 1 – Concept of information for use

Some product-specific information requirements (e.g. the wording of warnings or positioning of labels) are specified in standards for individual types or classes of products, but these do not provide a complete set of requirements for information for use. This document gives principles and general requirements for conveying information to users that are as applicable to complex and safety-critical systems (e.g. industrial plants), as they are to simple consumer products (e.g. a can of paint), to software, and to specialized testing equipment. Information for use is needed for anyone (skilled and unskilled) who encounters a product for the first time: whether to assemble from a kit, install, operate, maintain, or dispose of it.

The principles for preparing information for use of products are horizontally applicable across product sectors because all target audiences are human and subject to human error. The techniques found to be most effective to help such audiences to absorb new information are generally similar, as are their capabilities for misunderstanding language or images. What works best in information gathering and delivery (e.g. in content, wording, graphics, testing, and management of the whole process) has emerged from experience and practice in the fields of human factors and technical communication. This document is applicable on its own or can be referenced in product standards that include requirements to provide the target audience with information for use, for example, step-by-step instructions or other information products.

This document is addressed to those who prepare information for use; managers of organizations that produce or purchase products, systems, or services; human factors consultants; and product enforcement agencies.

It covers the following aspects:

- information content: conceptual information the target audiences need to understand, procedures they have to undertake, and the reference information they need to consult at some point;
- options for information to be provided as a single deliverable (e.g. a product manual) or across several types of information product, such as labelling on the product itself or packaging, accompanying electronic files, sheets, a website, booklets, printable files, videos, or searchable databases;
- effective use of language, text, illustrations, symbols, audio or video to communicate elements of information;
- processes and competencies involved in establishing content and preparing output; and
- means of assessing the fulfilment of requirements in accordance with this document.

PREPARATION OF INFORMATION FOR USE (INSTRUCTIONS FOR USE) OF PRODUCTS –

Part 1: Principles and general requirements

1 Scope

This part of 82079 International Standard provides principles and general requirements for information for the use of products.

Information for use is:

- necessary for the safe use of a product;
- helpful for the efficient and effective use of a product; and
- often necessary to fulfil market, legal, and regulatory obligations.

Products include, for example:

- industrial products (e.g. machinery, components, devices, and equipment);
- consumer products (e.g. household appliances, audio-visual devices, communication devices, and do-it-yourself products);
- medical devices, equipment and systems;
- complex systems of systems (e.g. industrial plants, refineries, production sites, and data centres);
- means of transport (e.g. cars, trucks, ships, and airplanes);
- application software (e.g. office software and web applications);
- software for operation and automatic control of systems; and
- technical services.

Information for use of products applies to phases of the product life cycle such as transport, assembly, installation, commissioning, operation, monitoring, troubleshooting, maintenance, repair, decommissioning, and disposal, and the appropriate tasks performed by skilled and unskilled persons.

This document provides the common and fundamental aspects serving as the binding and generic framework for prospective additional parts of this document.

This document applies to information for use whether provided as electronic or printed information products, for example:

- service information for machinery, provided as PDF file for web download for trained service technicians;
- information for operation of software, provided electronically with the software as an online help;
- troubleshooting information for an operator, on a machine's on-screen display;
- functional description of a medical device on a website;
- information for assembly, printed and provided in the packaging of a piece of furniture for consumers;
- printed information for maintenance for an automatic coffee machine;

- installation planning information for a safety sensor, downloadable from a website for mechanical engineers;
- product tutorial as a web-based training aid;
- materials for product on-site training; and
- label for transportation personnel, printed on the packaging of a heavy containment vessel.

This document is intended for use by all parties responsible for or involved in the conceptualization, creation, maintenance, translation, localization, integration of content, production, provision and evaluation, acquisition and supply of information for use.

The parties concerned with information for use include the following:

- acquirers and suppliers of products;
- managers with process or product responsibilities;
- content owners and content creators such as technical writers, information developers, and illustrators;
- technical translators, localization and terminology experts; and
- authorities, agencies and authorized experts.

The aim of this document is to provide these parties with the common and fundamental basis for developing information for use of supported products of the required quality.

This document is intended to be applied and referenced in product-specific standards, including those that specify the content of information for use of those products, for example, IEC 60335 for all parts for household electrical products, ISO 20607* for machinery, and ISO/IEC 26514 for systems and software. It is intended as a basis to elaborate product-specific requirements for target audiences or product information.

* Under preparation. Stage at the time of publication: ISO/DIS 20607:2018.

This horizontal standard is primarily intended for use by technical committees in the preparation of standards in accordance with the principles laid down in IEC Guide 108.

One of the responsibilities of a technical committee is, wherever applicable, to make use of horizontal standards in the preparation of its publications. The contents of this horizontal standard will not apply unless specifically referred to or included in the relevant publications.

2 Normative references

The following documents are referred to in the text in such a way that some or all of their content constitutes requirements of this document. For dated references, only the edition cited applies. For undated references, the latest edition of the referenced document (including any amendments) applies.

IEC 60417, *Graphical symbols for use on equipment* (available at http://www.graphical-symbols.info/equipment)

IEC 60617, *Graphical symbols for diagrams* (available at http://std.iec.ch/iec60617)

IEC 60825-1, *Safety of laser products – Part 1: Equipment classification and requirements*

ISO 3864 (all parts), *Graphical symbols – Safety colours and safety signs*

ISO 5807, *Information processing – Documentation symbols and conventions for data, program and system flowcharts, program network charts and system resources charts*

ISO 7000, *Graphical symbols for use on equipment* (available at http://www.graphical-symbols.info/equipment)

ISO 7010:2011, *Graphical symbols – Safety colours and safety signs – Registered safety signs*

ISO 9241-300, *Ergonomics of human-system interaction – Part 300: Introduction to electronic visual display requirements*

ISO 14617 (all parts), *Graphical symbols for diagrams*

3 Terms and definitions

For the purposes of this document, the following terms and definitions apply.

IEC, ISO, and IEEE maintain terminological databases for use in standardization at the following addresses:

- IEC Electropedia: available at http://www.electropedia.org/
- ISO Online browsing platform: available at http://www.iso.org/obp
- IEEE Standards Dictionary Online: available at http://ieeexplore.ieee.org/xpls/dictionary.jsp

3.1
accessibility
extent to which products, systems, services, environments and facilities can be used by people from a population with the widest range of user needs, characteristics, and capabilities to achieve identified goals in identified contexts of use

Note 1 to entry: Context of use includes direct use or use supported by assistive technologies.

[SOURCE: ISO 9241-940:2017, 3.1]

3.2
commissioning
procedures prior, or related, to the handing over of a product ready to be placed into service

Note 1 to entry: Commissioning may include final acceptance testing, the handing over of relevant documentation for the supported product or instructing personnel.

3.3
competent person
person who has acquired through training, qualifications or experience, or a combination of these, the knowledge and skills enabling that person to perform a specified task

Note 1 to entry: This document uses the term competent person in the context of the creation of the information for use and information management. See also definition 3.36, skilled person.

[SOURCE: ISO 17842-1:2015, 3.6, modified – Note 1 to entry added.]

3.4
component
product used as a constituent in an assembled product, system or plant

[SOURCE: IEC 81346-1:2009, 3.7]

3.5
consequence
outcome of an occurrence of a particular set of circumstances

Note 1 to entry: There can be more than one consequence from one occurrence.

Note 2 to entry: Consequences can be expressed qualitatively or quantitatively.

3.6
consumable
any part or material that is necessary to be replaced or refilled for continuous use or maintenance of the product

3.7
consumer
individual member of the general public, purchasing or using products for private purposes

Note 1 to entry: For the purpose of the requirements of this document, a "consumer" is assumed to be a non-skilled person.

3.8
consumer product
product available to, intended for or likely to be used by consumers

3.9
customer
individual or organization that purchases or receives a product

Note 1 to entry: The term "customer" includes but has a broader meaning than "consumer" (3.7).

3.10
document
fixed and structured amount of information intended for human perception that can be managed and interchanged as a unit between users and systems

[SOURCE: IEC 61082-1:2014, 3.1.1, modified – The notes have been deleted.]

3.11
documentation
collection of documents related to a given subject

[SOURCE: IEC 61082-1:2014, 3.1.3, modified – The notes have been deleted.]

3.12
equipment
associated assemblies intended to achieve a defined final objective

[SOURCE: IEC 80416-1:2008, 3.3]

3.13
graphical symbol
visually perceptible figure with a particular meaning used to transmit information independently of language

[SOURCE: IEC 80416-1:2008, 3.4]

3.14
harm
injury or damage to the health of people, or damage to property or the environment

[SOURCE: ISO/IEC Guide 51:2014, 3.1]

3.15
hazard
potential source of harm

[SOURCE: ISO/IEC Guide 51:2014, 3.2]

3.16
illustration
visually perceptible figure, artificially created to transmit specific information, excluding safety signs and graphical symbols

EXAMPLE Graphical image, picture, figure, photograph, line drawing, table.

3.17
information for use
instructions for use
information provided by the supplier that provides the target audience with concepts, procedures and reference material for the safe, effective, and efficient use of a supported product during its life cycle

EXAMPLE Step-by-step instructions, troubleshooting information, service information, operation and maintenance instructions, and assembly instructions.

Note 1 to entry: "Instructions for use, procedures for the safe, effective, and efficient use of a supported product", was the term formerly used to include all the content defined as part of information for use.

Note 2 to entry: Excludes supplementary information, which is outside the scope of information for use.

3.18
information model
representation of concepts, relationships and rules for a body of information

3.19
information product
information item
separately identifiable body of information that is produced, stored, and delivered for human use

Note 1 to entry: A document produced to meet information requirements can be an information product, or part of an information product, or a combination of several information products.

Note 2 to entry: An information product can be produced in several versions during a project or system life cycle.

[SOURCE: ISO/IEC/IEEE 15289:2017, 3.1.13, modified – "information product" has been added as a first preferred term and in the notes "information item" has been replaced with "information product".]

3.20
information quality
degree to which an information product satisfies the stated and implied needs when used under specified conditions

3.21
intended use
exhaustive range of functions or foreseen applications defined and designed by the supplier of the product

Note 1 to entry: Functions or applications not listed by the supplier are excluded from the intended use of the product.

Note 2 to entry: Additional or modified functions or applications resulting from modifications not sanctioned by the supplier of the product are excluded from the intended use.

3.22
label
item, attached to a product (if practicable) or its packaging, which displays information related to one or more characteristic(s) of the product

[SOURCE: ISO/IEC Guide 14:2018, 3.10, modified – changed "consumer product" to "product".]

3.23
maintenance
actions intended to retain a product in, or restore it to, a useful and safe condition, in which it can perform the intended use

3.24
marking
symbols, pictograms, warnings, logos or inscriptions on the product, label, or packaging to identify its type, which can also include short textual messages

[SOURCE: ISO/IEC Guide 14:2018, 3.12, modified – changed "consumer product" to "product".]

3.25
minimalism
principle that information for use includes critical information and the least amount of other information needed to be complete

3.26
personal protective equipment
any device or appliance designed to be worn or held by an individual for protection against one or more health and safety hazards

[SOURCE: IEC 60050-651:2014, 651-23-01, modified – The abbreviated term "PPE", the wording "whilst performing live working" and Note 1 to entry have been deleted.]

3.27
plant
assembly of different systems on a specific site

[SOURCE: IEC 61355-1:2008, 3.10]

3.28
product
intended or accomplished result of labour, or of a natural or artificial process

Note 1 to entry: Information for use is seen as a part of a product.

Note 2 to entry: A product can be a system, a service, goods, information, or a combination thereof.

3.29
product safety label
label on a product that informs of one or more potential hazards and describes the safety precautions and/or actions required to avoid the hazard(s)

[SOURCE: ISO 3864-2:2016, 3.9, modified –"the observer" has been deleted and Note 1 to entry has been deleted.]

3.30
reasonably foreseeable misuse
use of a product or system in a way not intended by the supplier, but which can result from readily predictable human behaviour

Note 1 to entry: Readily predictable human behaviour includes the behaviour of all types of users. For more information, see ISO 10377.

[SOURCE: ISO/IEC Guide 51:2014, 3.7, modified – The wording "e.g. the elderly, children and persons with disabilities" has been deleted from Note 1 to entry. Note 2 to entry has been deleted.]

3.31
repair
corrective maintenance of defective or damaged parts or functions of a product

3.32
risk
combination of the probability of occurrence of harm and the severity of that harm

Note 1 to entry: The probability of occurrence includes the exposure to a hazardous situation, the occurrence of a hazardous event and the possibility to avoid or limit the harm.

[SOURCE: ISO/IEC Guide 51:2014, 3.9]

3.33
safety note
safety-related information that is collected or grouped in a document or section of a document in a meaningful organizational system to explain safety measures, raise safety awareness, and provide a basis for safety-related training of persons

3.34
safety sign
sign giving a general safety message, obtained by a combination of colour and geometric shape and which, by the addition of a graphical symbol, gives a particular safety message

[SOURCE: ISO 7010:2011, 3.3]

3.35
self-assembly product
flatpack product
ready-to-assemble product
product that is intended to be assembled or erected from separate components without the help of a supervisor or trainer and relying only on step-by-step instructions

Note 1 to entry: This includes both products supplied in the form of a kit of component parts and products that incorporate structural elements requiring folding, locking or tensioning (e.g. child pushchairs or camping equipment).

3.36
skilled person
individual with relevant technical education, training, or experience to enable perceiving risks and avoiding hazards occurring during use of a product

3.37
step
one element in a procedure that tells the target audience to perform an action

Note 1 to entry: A step is presented as a numbered list item.

3.38
step-by-step instructions
procedure consisting of steps to be followed for the safe, effective, and efficient use of a supported product

3.39
supplier
individual or organization that provides products

Note 1 to entry: "Supplier" includes producer, manufacturer, integrator, vendor, distributor, retailer, agency, provider of information or their authorized representatives.

3.40
supported product
product to which the information for use relates

3.41
system
set of interrelated elements considered in a defined context as a whole and separated from their environment

[SOURCE: IEC 60050-151:2001, 151-11-27, modified – The notes have been deleted.]

3.42
target audience
group of persons for whom information for use is intended by the supplier

3.43
topic
coherent part of an information product with a heading that deals with a single subject

Note 1 to entry: In a printed document a topic is a section or subsection. In online documentation, a topic usually is the amount of information invoked by navigation (e.g. following a link) and completely accessible by scrolling.

3.44
usability
extent to which a product can be used by specified users to achieve specified goals with effectiveness, efficiency, and satisfaction in a specified context of use

[SOURCE: ISO 9241-210:2010, 2.13 modified – "system or service" have been deleted.]

3.45
use
activity that the user may perform with or on the product during its whole life cycle

Note 1 to entry: "Use" covers the intended use and the reasonably foreseeable misuse in normal and reasonably foreseeable conditions of use.

3.46
use case
description of behavioural requirements of a system and its interaction with a user

[SOURCE: ISO/IEC/IEEE 26515:2018, 4.15]

3.47
user
person who interacts with the product

Note 1 to entry: "User" can include persons who install, operate, service, maintain, or dispose of the product.

3.48
warning message
safety-related information that warns about hazards and instructs on how to avoid them

Note 1 to entry: Warning messages are normally given within step-by-step instructions related to hazardous tasks.

4 Fulfilment of requirements for information for use

4.1 General

The fulfilment of specified requirements in this document does not release suppliers from their obligations to assess the legal framework for product compliance within each separate distribution area. Inadequate product-related information can result in product safety and liability risks.

The requirements of this document are of two kinds:

a) requirements for information for use supporting a particular product in 5.2, 5.3, Clause 7, Clause 8 and Clause 9; and

b) requirements for information management processes in 5.4, Clause 6, and Clause 10.

A claim of fulfilment of requirements for information for use of a particular consumer product shall be made by a supplier only if the information products supplied (including information products supplied electronically) to support the product fulfil the requirements in a).

Claims of fulfilment of requirements other than for information for use supporting a consumer product shall only be made by a supplier if:

1) the processes established by an organization to prepare information for use fulfil b); or

2) the information for use supplied to support a particular product fulfil a).

A supplier may claim fulfilment of requirements for both the information for use supporting a particular product (a)) and for the information management process (b)).

Annex A provides guidance on evaluation.

4.2 Evaluation of information for use of consumer products

Evaluation of information for use of a consumer product should include an empirical effectiveness check. The empirical effectiveness check should involve not only skilled experts but also persons unfamiliar with the use of the supported product and its information for use. Where practicable, the supported product and preferably packaging should be available for the evaluation. This should be practicable for consumer products.

Annex A describes several methods for empirical effectiveness checks of information products.

4.3 Documentary evidence of evaluation

Where fulfilment of requirements in this document is claimed, the supplier shall retain records of the following:

a) the information for use and any packaging that were submitted for the evaluation;

b) the language version of the information for use which was assessed and approved as fulfiling the requirements in this document;

c) whether the product sample was supplied fully assembled or was assembled as part of the assessment of the information for use; and

d) the name of the organization or person(s) that carried out the evaluation.

5 Principles

5.1 General

Clause 5 establishes principles for the selection of content, structure, and format for information for use, and the processes for the design and sustainment of information for use. The principles are concerned with:

a) the purpose of the information for use;

b) the quality of information in the information for use; and

c) the use of processes for the management of information.

5.2 Purpose of information for use

5.2.1 General

Information for use shall provide the necessary information for the target audiences to make safe, efficient, and effective use of a supported product.

To meet this purpose, information for use provides the following information types:

a) conceptual information: concepts, explanations and descriptions to enable the target audiences to perform tasks by understanding their purpose and the principles of operation of the supported product;

b) instructional information: procedures and task-oriented step-by-step instructions; and

c) reference information: detailed information which needs to be retrieved occasionally, such as troubleshooting information, commands or codes.

5.2.2 Information for use as part of the product

Information for use is an integral part of the supported product and shall be given the same attention and importance as every other part of the product.

If the information for use is defective (e.g. does not meet the requirements), the product as a whole is defective. Information for use shall be clearly and easily identifiable with the supported product.

5.2.3 Target audiences orientation

Information for use shall be usable and relevant for the target audiences with respect to their expected tasks and goals.

NOTE More information on target audience analysis can be found in ISO/IEC 26514.

5.2.4 Safe use of the supported product

Information for use shall promote the safe operation and maintenance of the supported product. Information for use contributes to reducing the risk of injury or illness to people or animals, and risks of damage, malfunction or inefficient operation of the product. However, information for use cannot compensate for design deficiencies or missing safety systems.

5.2.5 Product compliance through information for use

The framework for product compliance consists, in general, of the following areas of legal practice:

a) product safety;

b) product liability; and

c) contracts.

Because the information for use is part of the supported product, the legal compliance of the information for use can be considered as part of the legal compliance of the product.

5.3 Information quality

5.3.1 General

Information for use shall meet the target audiences' needs for information quality: completeness, minimalism, correctness, conciseness, consistency, comprehensibility, and accessibility.

Information quality means that the information meets the requirements of all parties with an interest in the supported product with respect to the content; structure; means, format and media of information.

NOTE Poor information quality can cause legal defects.

5.3.2 Completeness

Information for use shall be complete. The completeness of information depends on the following being addressed:

a) the risks of using the product;

b) the tasks that the target audiences are intended or allowed to perform throughout the life cycle of the supported product;

c) the target audiences' need for information; and

d) legal and contractual requirements.

5.3.3 Minimalism

Minimalism shall be applied. Minimalism is an approach to information for use that includes critical information and the least amount of other information needed to be complete. Critical information includes the safe use of the product, the security of the information created with the product, or the privacy of the information created by or stored with the product.

NOTE Minimalism means that only relevant information is provided.

In contrast to minimalist information, excessive documentation of all possible product functions, for example, is wasteful. It can hinder users' comprehension of how to use a product for their common or critical tasks, and from discovering needed information quickly in their context of use. Minimalism balances the cost of developing and sustaining information for use with the needs of the target audiences to operate a product safely, effectively, and efficiently.

Critical information, such as safety-related information, shall be repeated where it is needed.

5.3.4 Correctness

Information for use shall be technically correct and contain the current information on the supported product to which the information for use relates. If the supported product is being modified, the need for changes in the information for use should be evaluated.

5.3.5 Conciseness

Information for use shall be concise with respect to format and media.

Conciseness with respect to the content and presentation of the information means:

a) succinct wording;
b) text and illustrations without unnecessary details; and
c) videos as short as relevant.

5.3.6 Consistency

Information for use shall be consistent with respect to the content, format and media.

Consistency with respect to content means, for example, that the information is unambiguous and correct (e.g. consistent with the supported product). Consistency with respect to format and media means, for example, consistent terminology, units of measurement, structures and indications (i.e. headings, titles), symbols, warning messages, and use of colours.

To achieve consistency, systems for consistent formatting should be separated from systems for consistent content (presentation of topics). This is called the principle of separation of content and format.

NOTE Mark-up languages can help to achieve consistency.

Information for use should be consistent with other information for the supported product, such as that on or in product packaging, labelling, training materials, materials for customer support and user assistance, and promotional materials.

5.3.7 Comprehensibility

Information for use shall be comprehensible to the target audiences. This requires:

a) comprehensible text and terminology;
b) comprehensible illustrations, safety signs and graphical symbols; and
c) comprehensible methods of navigation and use of media.

NOTE Comprehensibility includes clarity and intelligibility. Comprehensibility depends on the context of use. The need for comprehensibility is especially important in emergency situations.

5.3.8 Accessibility

Information for use shall be accessible (e.g. technically available and legible) for the target audiences in the context of use and throughout the intended lifetime of the supported product.

5.4 Use of repeatable processes

Information for use shall be developed, produced, and sustained using defined, repeatable processes. The use of repeatable processes facilitates planning, implementing, checking, correcting, and improving the information for use.

These processes shall:

a) be planned so as to obtain sufficient competent human resources for information development and to make information accessible at the required time;
b) be quality managed and assured on the basis of clear and common quality objectives; and
c) use single-source information management to reduce inconsistencies and discrepancies.

Information content should be retained for reuse throughout the intended lifetime of the supported product, maintained to be current, and adaptable for use on similar products and future product versions.

6 Information management process

6.1 General

The supplier shall implement information management processes for planning, designing, producing, and sustaining information for use. The information management process shall generate, obtain, confirm, transform, retain, retrieve, and disseminate information, as appropriate.

The information management process shall:

a) identify information to be managed;

b) define information representations;

c) obtain, develop or transform, store, validate, present, and dispose of information;

d) identify the status of information; and

e) make information available for use.

The development and sustainment of information for use are typically performed as a supporting information management process within the life cycle of the supported product.

Four process groups shall be addressed:

1) analysis and planning of information (6.2);

2) design and development, including review, editing, and testing (6.3);

3) production and distribution (6.4); and

4) sustainment, including maintenance and improvement (6.5).

6.2 Analysis and planning of information

6.2.1 General analysis

Analysis and planning of information shall begin with an organizational strategy and policy regarding the development and sustainment of information for use. Senior management should arrange the resources, schedules, and quality targets so as to enable the planning, preparation and delivery of information for use.

Instructional procedures shall be formulated based on a market analysis or an analysis of the characteristics, needs and intended tasks of the target audiences. The analyses shall be performed early in the life cycle of the supported product, and by collecting information about the product. This analysis should result in an understanding of the context of use (operational concept) and the nature of the target audiences. Decisions made during analysis should include the following:

a) the type of media to be used;

b) whether to segment information for various target audiences or language versions;

c) whether to include all information at one website, or in one document, or a series of documents; and

d) whether to include the information within or on the supported product or provide the information separately.

Factors affecting the information design for information for use shall include target audience analysis, media and languages, information sources, risk management, and legal considerations.

6.2.2 Target audiences

Planning for information for use shall consider the skills and tasks required of the target audiences. In this context, there can be multiple target audiences, including users of the supported product, installers, technicians, administrators, operators, and those responsible for security, product disposal or recycling. The expected degree of knowledge and skills of the target audiences, and their expected frequency of use of the supported product should be defined.

Planning for information for use shall consider the characteristics of the target audiences. Each target audience should be defined by common characteristics, for example, similar work tasks.

The definition for each target audience should consider the following characteristics:

a) background, skills, experience, and training with respect to the relevant sectors, for example, mechanics, electrics, electronics, hydraulics, software, programming;

b) language familiar to the target audience;

c) tasks;

d) working environment, for example, climatic conditions, light, working in a team or alone;

e) available tools; and

f) available means for information access, for example, Internet access, audio services, phone or video remote services to get support, permission for information access, adequate device for displaying electronic information.

The description of target audiences should consider that target audiences are often heterogeneous, i.e. including a range of capabilities.

The description of the tasks of the particular target audiences should consider the following aspects, if applicable:

1) why the task is carried out;

2) how frequently the task is carried out and how likely it is that the users will remember how to do it;

3) how long the task is expected to take, for example, minutes, hours, or days;

4) whether the task is carried out by one person, or is carried out as a process by more persons with one or more different job roles;

5) what the specific environmental conditions of the task are, for example, stress and mental pressure, climatic conditions;

6) how fault-tolerant the task is and how important it is that the user carries out the task correctly;

7) what discretion the user has in how or when the task is carried out;

8) what the prerequisites for the task are; and

9) how common the task is.

NOTE For further information on target audience analysis and the method of audience mapping matrix, see ISO/IEC 26514.

Target audiences with particular needs for accessible information include older persons; children whose use of the supported product requires supervision; persons unable to read the printed information for use; and persons with physical, visual or auditory disabilities.

6.2.3 Media

Analysis and planning of information shall include the choice of media for the information for use, such as electronic or printed media, provided as part of the supported product, its packaging, or separately. The supplier should consider the choice of media in view of the needs of the target audience for information accessibility. The produced (for example, printed) information for use should be usable for the intended lifetime of the supported product; for example, printed information for use should be provided on durable paper. Electronic information for use can be provided in combinations of audio, visual (i.e. illustration, video, or animation), or written forms, on separate media, through the Internet, for use on various stationary or mobile viewing devices, or embedded in the supported product.

6.2.4 Languages

Planning of information shall include identification of local language requirements for the information for use. These shall be identified and versions of the information for use shall be translated and provided accordingly.

NOTE 1 For further information on the translation management process, see ISO 17100.

Consideration should be given to conveying important installation and operational precautions and safety-related information by means of language-independent safety signs, graphical symbols and self-explanatory illustrations.

NOTE 2 It is usual for information for use to be supplied in the official language(s) of the country of sale.

6.2.5 Information sources

Planning of information shall include the identification of information sources and the most effective ways to use them. The following should be taken into account:

a) function, structure and life cycle of the supported product, similar products and requirements for information for use, possible problems and troubleshooting information, spare parts and options, compliance and conformance information;
b) the market(s) in which the supported product will be supplied, including characteristics of the market(s) and channels for distribution of the supported product;
c) legal and normative requirements for the markets in which the supported products will be supplied; and
d) the time when stable and reliable information will be available.

Sources for information for use should include interviews and observations by persons having experience with the supported product. Where information for use integrates several already documented systems or services as information sources, the supplier should decide whether to adapt the existing documentation using its own formats or incorporate the existing documentation by reference without change, using cross-references or hyperlinks.

6.2.6 Information sustainment

The needs for maintaining and supporting information during the life cycle of the supported product shall be taken into account when planning the preparation of information for use. Information shall be managed so that relevant information on product changes and new releases is available for inclusion in the process. Planning for information for use shall include identification of the expected duration and level of post-delivery support for the supported product and, therefore for the information for use, along with plans to store, retrieve, change, protect, and dispose of the information for use.

6.2.7 Risk management

Analysis and planning of information shall be based on considerations of hazards inherent in the installation, operation, and maintenance of the supported product. The analysis of hazards shall consider tasks or use cases. Hazards include not only immediate risks to operators and users of the supported product, but also foreseeable future impact on others in the environment. Hazards also include the possibility of unfavourable or unexpected risks for the customer, such as loss of business or personal data. The supplier shall identify, analyse, and evaluate hazards for their impact, likelihood of occurrence, and availability of methods to avoid or limit them. Residual risks (risk remaining after risk reduction) shall be included in the information for use as limitations or special exceptions and addressed in warning messages.

The planning process shall include the treatment of risks in the information for use. This should be coordinated with the engineering and design functions to help ensure the safe intended use of the supported product. With respect to safety, the developer of the information for use should be involved in risk assessments so as to make clear the limits of information for use in ensuring safety, and to help build safety into the design of the supported product through inherent safety features, auxiliary safety devices, safety warnings in user interface messages, safety notes, training, and packaging.

NOTE ISO 12100, ISO 31000, ISO 14971 and ISO/IEC Guide 51 provide additional requirements for risk management and safety-related information.

6.2.8 Contractual agreements and legal constraints

Requirements regarding information for use resulting from contractual agreements and legal constraints shall be available for the information development process.

6.2.9 Project management and control

Managing projects shall include planning, developing, maintaining project schedules, performing resource management, and measuring and reporting the status of the project and the information for use. Planning should include the following:

a) adoption of policies for content management and reusability;

b) review and testing;

c) preparation of translated or localized versions;

d) resource management;

e) the involvement of competent persons;

f) quality assurance, project risk management, and configuration management;

g) obtaining infrastructure resources, such as a content management system and publishing system, or the use of an information management system integrated with the project's software development tools;

h) management of the communication within the project team, with information providers and with stakeholders; and

i) for complex systems, a delivery plan for the different information products.

Control shall include the responsibility for authorizing and closing projects, obtaining and communicating accurate and timely information on project status, and redirecting project activities and tasks to take advantage of opportunities or to correct deficiencies.

NOTE Information about management for information development can be found in ISO/IEC/IEEE 26511.

6.2.10 Configuration management

Defective configuration management can result in wrong or missing information in the information for use for a specific product. Persons in charge of configuration management shall establish baseline versions and manage changes to information for use under their control. They shall control work information, such as reusable content ("single source") for information for use, product components, plans and design artefacts. They should manage traceability between versions of information for use and supported products, so that it is possible to find the current or applicable version of the information for use at any given time after release throughout the life cycle of the supported product. They shall include activities to review and approve changes and releases.

NOTE ISO/IEC/IEEE 26531 includes requirements for content management and a component content management system.

6.2.11 Human resource management

Preparation of information for use shall include planning for, obtaining, and allocating the required competent human resources. This shall include training in the use of software tools and methods and familiarization with organizational procedures where necessary.

6.2.12 Quality assurance

When preparing information for use, quality assurance techniques and activities shall be identified and performed throughout the information development process.

NOTE ISO 9000 describes quality management terms and concepts.

6.3 Design and development, including review, editing, and testing

6.3.1 General

The development process shall include:

a) preparing information design concepts, templates or models;

b) gathering information;

c) selecting the content structure;

d) applying the information design to create the written and graphic content;

e) implementing the information in the specified media; and

f) editing and reviewing the content, and evaluating the information for use with the supported product.

Usability testing should be used to demonstrate that information for use allows the target audiences to perform the described tasks, or to point out defects.

6.3.2 Design, information gathering and development

A consistent design concept should be used for related information for use. A style guide or guideline for authoring and editing should be established and followed throughout the information for use, covering, for example:

- writing style;
- wording;
- consistent use of terms;
- form of addressing or referring to the target audiences;
- design of text and page layout including selection of typeface and font size;
- navigation and linking; and
- use of the structures for information types (e.g. safety notes and procedures).

Consistent terminology should be agreed, recorded, and communicated for use throughout the information for use. If translation is required, the translation provider should provide information showing how terms have been translated consistently for the target language.

NOTE For further information on terminology work see ISO 704.

Empirical tests (see Table 1 in 6.3.3) should be carried out if there are any doubts that technical terms will be understood by the target audience.

The underlying terminology should be defined and documented before drafting the content of the information for use.

6.3.3 Reviewing, editing and testing

Reviewing and testing of the information for use should be performed throughout the information development process.

Evaluations should be performed using methods such as peer reviews, walkthroughs and inspections by experts outside the team responsible for development of the supported product. Interactive empirical evaluation of the information for use should be carried out by a panel representing the range of target audiences. Informal or controlled usability testing should also be carried out. Editing should be performed by persons other than those involved in writing or translating the information for use.

The source text should be reviewed for accuracy, consistency, and usability before it is translated. Translations should be reviewed for technical and linguistic accuracy and clarity by competent persons; details are given in 10.4.

Empirical evaluation (e.g. usability testing) should be used as a means for determining whether the information for use achieves its purpose. Table 1 shows examples of empirical methods.

Table 1 – Examples of empirical methods

Empirical method	Scope and objectives of the method
Interviews for opinion survey	analysis of satisfaction by those involved in the preparation or presentation of the information for use target audience analysis identification of information regarding the context of use of the product (for example why, where, how often) identification of target audiences' needs assessment or evaluation of information for use comparison of different information products
Focus groups for opinion survey	identification of target audiences' needs assessment or evaluation regarding handling of and working with the information for use comparison of different styles and formats
Usability testing	identification of gaps and weaknesses in the information for use testing of information for use or comparison of different versions testing of comprehensibility and usability
Self-assessments and checklists	assessment of quality according to technical criteria checklists relating to standards
Expertise, expert reviews and certification	assessment of information for use according to defined criteria analysis of weak points attention to legal concerns
Feedback, recommendations, or observations from independent award or merit schemes	evaluation of a product according to generally accepted guidelines comparison with other information products

Empirical method	Scope and objectives of the method
Feedback from complaints management, hotlines and service information	continuing improvement process optimization of request handling improvement of the information product

Improvements should be incorporated into the style guide in order to continually improve the quality of information for use. Errors discovered during reviews and testing shall be recorded, analysed for severity and time and cost to correct, and resolved.

NOTE 1 For example, a minor spelling error can be immediately corrected, but an error affecting safety-critical systems could cause a project shutdown until the problem is resolved.

Project schedules should allow time for reviews and correction of significant defects, with repeated reviews if needed. A designated person shall be the review authority for determining whether the information for use has met its review criteria and can proceed to the next stage.

NOTE 2 ISO/IEC/IEEE 26513 provides detailed requirements for review and test of information for users.

6.4 Production and distribution

Production of information for use shall comprise the integration, preparation, reproduction, packaging and distribution of physical media or electronic copies of the information for use. Suppliers should identify the information for use for the supported product in their logistics and order processing systems.

6.5 Sustainment, maintenance and improvement

Sustainment shall include continuing target audience feedback, in order to keep the information for use current for different versions of a supported product throughout the life cycle of those products, including feedback on modifications relating to improved usability.

The supplier shall establish a method for receiving information on changes, updating information for use, and making updates available to the target audiences.

Information on customer satisfaction should be used to amend information for use as appropriate. Such information can be found from customer complaints, service reports, analyses of causes of accidents and injuries, and reviews of products posted on the Internet.

After the sale of the product, the target audiences should be informed of safety-critical revisions to information for use via websites, e-mails, or mass media announcements.

Suppliers should encourage target audiences to provide feedback on experiences with using the information for use and the supported products.

Suppliers can consider related information published by third parties to identify improvements or missing content in the suppliers' information for use.

7 Content of information for use

7.1 General

The order of subclauses in Clauses 7 and 8 does not imply that the information for use should be developed in this order or presented to the target audiences in this order.

The content of the information for use shall cover the needs of the target audiences for the safe, effective and efficient use of the product, applying the principle of minimalism.

Information for use shall describe the intended use of the product and provide information that helps the target audiences avoid unacceptable safety risks, malfunction of or damage to the supported product, or inefficient operation. Information for use shall provide the target audiences with the necessary information to identify and avoid reasonably foreseeable misuse of the product as identified in accordance with 6.2.7.

7.2 Identifiers

7.2.1 Identification of information for use

Information for use can be supplied through a set of information products. Each information product shall be unambiguously identifiable. It shall be designated by a unique identifier and reference information, including, where applicable:

a) title of the information product;

b) identification number;

c) date of issue;

d) revision number and date of issue of the revision of the information product, where applicable; and

e) identifier for the language.

NOTE ISO 639-2 provides standard codes for the names of languages, which can identify the various languages in information for use.

7.2.2 Identification of the supported product

Information for use shall clearly identify the supported product. The information for use shall display the unique identifier of the supported product, including, where applicable:

a) name;

b) serial number, type number or lot number;

c) version number;

d) information about the configuration; and

e) illustrations.

Where the product is in conformance with an international standard, this should be indicated. Where more than one product version is addressed by the information for use, the identifier of a specific version shall be clearly recognizable in the information for use and on the product, to facilitate identification. The relationship between content and operational features of different product variants should be unambiguous.

NOTE For information about identification requirements, see IEC 62507-1.

7.2.3 Identification of the supplier

Information for use shall indicate the supplier of the supported product. It shall therefore display:

a) name and address;

b) contact details such as telephone numbers, email address, hyperlinks and other possible means of communication; and

c) contact details of providers of special tools, material, authorized service centres for the product and technical assistance, if offered.

7.3 Importance of retaining printed information for use

If the information for use is delivered in paper form, the target audience's attention shall be drawn to the importance of retaining it for the intended lifetime of the supported product by a statement as shown by the two examples in Figure 2:

Example 1:

IMPORTANT
Read carefully before use.
Keep for future reference.

Example 2:

For safe and proper use,
follow these instructions.
Keep them for future reference.

IEC

Figure 2 – Examples for notice to retain information

NOTE For the safety sign, see ISO 7010-M002:2011-05.

Replacement copies of information for use shall be made available from the supplier (for example, from a website) throughout the intended lifetime of the supported product.

7.4 Presentational conventions

Presentational conventions used in information for use shall be listed and explained when this convention is essential to understand the information, for example, the use of a particular typeface for text to be entered by the user, keyboard keys, or menu items.

7.5 Terminology

Consistent terminology shall be used throughout the information for use, on the supported product, and on its packaging. Ambiguities shall be avoided. Colloquial expressions and untypical regional variations of names and product features should be avoided.

Unless the supported product is intended for use by a target audience with specialist knowledge, terminology shall be explained, for example by adding definitions, links, references or glossaries.

If the information for use of a system contains documents from various other suppliers, consistency of all terms in all documents cannot always be achieved. In this case the supplier shall explain the meaning of the differing terms so that the target audience can easily relate the different terms in different documents.

7.6 Acronyms, abbreviations and technical terms

Unavoidable acronyms, abbreviations and technical terms not readily understood by the target audience shall be listed and explained.

7.7 Explanation of safety signs, graphical symbols and markings

Safety signs, graphical symbols, pictograms, and markings placed on the supported product and packaging shall be explained.

7.8 Product description

7.8.1 General description

Information for use shall contain a general description of the supported product so that the target audience can understand the essential functions of the product in the contexts in which it is intended to be used. The intended uses of products shall be clearly and unambiguously described, including the technical limits and conditions within which the product is intended to be used.

7.8.2 Visualization

In addition to the general description, a visualization of the supported product should be provided to help the target audience identify the product and its main functions.

7.8.3 Specifications

The product description shall include the following specifications where applicable:

a) complete measurements, for example, of mass, volumetric capacity, and performance;

b) information regarding electrical ratings (e.g. input voltage);

c) information regarding energy consumption, energy efficiency rating, insulation category, and IP code;

NOTE IP Code is defined in IEC 60529.

d) communication interfaces and protocols;

e) gas supply (for example, type, pressure); water supply and other relevant information on items such as cleaning agents, lubricants, and fuses (for example, type, rated value and characteristics);

f) by-products of use under specified circumstances, such as emitted noise level, waste discharge rate;

g) electromagnetic compatibility (EMC);

h) description of the type of intended user and, especially in the case of accessible design products, a clear statement of any special precautions that should be followed by the user;

i) standards, and legal and regulatory obligations with which the supported product complies;

j) notes on intended use and reasonably foreseeable misuse of the supported product;

k) description of the user controls and indicators applicable to the relevant component(s);

l) acceptable environmental conditions, such as ambient temperature or humidity; and

m) information about the year of manufacture, expiration date, and limitations on hours of use, if supported products have an intended limited lifetime, for example, for safety reasons, quality or economic performance.

For principles and methods for detailed specification of products, see IEC 62569-1.

7.9 Supplied accessories, consumables and spare parts

7.9.1 Accessories supplied with the product

The following information, as required for each supplied accessory, shall be included in the information for use:

a) name, part number;

b) illustrations to allow the target audiences to identify the accessory and where to attach it;

c) description of its function and purpose;

d) information needed to install the accessory, for example with an illustration, and how to use it; and

e) information needed to order a replacement.

7.9.2 Consumables

Where consumables are required, the following information shall be provided in the information for use:

a) description and recommended quantity of consumables and where applicable, the supplier's model name or part number, and a generic name;

b) an illustration showing the position of, or access to, the consumables (preferably in an exploded view);

c) information for removal and replacement or refilling;

d) information about disposal of consumables; and

e) information on when a consumable is no longer usable, for example, signs of material degradation or excessive storage time or number of allowable uses.

7.9.3 Spare parts

For products with replaceable parts, a list of spare parts shall be provided. The following information should, as appropriate, be provided for each spare part:

a) name, identification and serial number in the form used on the supported product;

b) supplier's part or type and version numbers as well as the name, reference and version numbers for branded parts, if different from the above;

c) illustration of the part and its position, preferably in exploded view;

d) sources of supply as well as alternative sources of supply for spare parts;

e) parts for which there is a refurbishing service available provided by the supplier or a service provider;

f) for industrial systems, identification according to IEC 81346-1 for components and locations where a spare part can be used as a replacement;

g) the year of manufacture or the year for the availability of the spare part if not given in a separate document; and

h) information about disposal of spare parts.

7.10 Information for use needed during the lifetime of the supported product

7.10.1 General

Where legal requirements, safety, or technical reasons prohibit the installation, operation, or maintenance of the supported product by non-skilled persons, this shall be made clear in the information for use. In such cases, the qualifications, competencies, and skills required shall be clearly stated in the information for use.

7.10.2 Re-packaging of the supported product

Where applicable, information for use shall include information on special precautions to be taken when repackaging.

7.10.3 Transportation and storage of the supported product

Where applicable, information for use shall include information for the protection of the supported product and the safety of persons during transportation and storage, for example:

- dimensions, mass, and centre of gravity;

- methods of lifting, handling and transportation (e.g., by indicating gripping points and lock-down screws) that can prevent damaging impacts and can help avoid strain injury to the user;
- storage requirements with associated ambient conditions, for example, humidity, storing flat, avoiding direct sunlight;
- storage of the supported product during transport before and after use; and
- methods for securing the supported product against unexpected shocks.

7.10.4 Installation of the supported product

Where applicable, information for use shall include information on installation, as follows:

a) what to do if the product itself or the packaging is damaged or unintentionally opened before use;

b) methods for the removal of transport packaging restraints and transit clamps;

c) procedures for unpacking (provided on the outside of the packaging) and for the removal and safe disposal of protective and preservative packaging;

d) a checklist of items included in the package;

e) minimum space needed for use, maintenance and repair;

f) a layout plan;

g) an interconnection diagram or table;

h) conditions for assembly and installation including earthquake-resistant considerations;

NOTE For self-assembly products, see 7.12.

i) methods, precautions and legal restrictions for connecting the supported product to power and water supplies, drainage and other auxiliary supplies; and

j) checking whether the product is properly installed and ready to perform safely and as intended by the supplier.

If a product is supplied to locations that prohibit the installation by non-skilled persons, the information for use shall state that installation requires special expertise and specify the needed expertise.

7.10.5 Commissioning of the supported product

Information for use shall include the information necessary for the commissioning of the supported product prior to its handing over in working order.

7.10.6 Modification of the supported product

Where modification of a supported product is not permitted, this restriction shall be clearly stated in the information for use, together with information about possible consequences resulting from a modification. These consequences can include invalidation of warranties and access to support services.

Where modification of a supported product is permitted, the information for use shall clearly describe and illustrate how to undertake such modifications or direct the target audiences to a source of relevant information.

7.10.7 Operation of the supported product

7.10.7.1 General

Information for use shall provide information on the following:

a) normal operation; and

b) exceptional and emergency situations.

7.10.7.2 Normal operation

Information for use shall, where applicable, include the following:

a) illustrations that clarify or reinforce understanding of the main functions and safety precautions;
b) information related to when and how the supported product should be used;
c) information about local or remote operation and how to switch between the two;
d) information about manual operation and automatic operation and how to switch between the modes of operation;
e) means of starting and stopping the supported product's operation;
f) initiating operations, checks, adjustments, or conditioning required prior to each occasion of normal use;
g) identification and correction of malfunctions; and
h) recommendations for waste disposal.

Information regarding operational control shall explain indicators of the operational state as appropriate regarding:

1) normal operation;
2) indications that the system has automatically switched to safe operation mode before a hazardous situation has been reached;
3) the current state of operation;
4) what to do when the control system itself has failed; and
5) whether the system is in manual, local, or automatic mode.

NOTE For software documentation for user guidance systems, refer to IEC 61506.

7.10.7.3 Exceptional and emergency situations

Information for use shall, where applicable, cover the following:

a) sequence of actions to be taken in the case of an unexpected interruption of power, cooling, control functions, or another emergency situation;
b) fault indication and its location;
c) restarting after an intervention;
d) the type of firefighting equipment to be used in the case of fire;
e) foreseeable emissions, leaks, or decomposition of harmful substances, together with associated warnings of the likely hazards and precautions for avoiding harm; and
f) first aid treatments recommended for foreseeable acute medical conditions that can result from the use of the supported product, for example, toxic shock, epileptic seizure or burns, and advice on subsequent professional medical treatment to be sought.

7.10.8 Indications of faults and warning device signals of the supported product

Recognizable indications of normal and abnormal operation and the recognition of signals from any warning devices shall be explained in instructional information that is accessible without the operation of the supported product.

Signals or displays intended to indicate impending risk of damage or imminent danger shall be readily comprehensible and unambiguous.

7.10.9 Meanings of signals of the supported product

The meanings of signals, such as flashing indicators, colour changes or audible signals shall be explained so that they are easily recognized and understood by the target audience.

NOTE Examples of signal signs are found in IEC 60073, ISO 7731, ISO 11429, ISO 15006 and ISO TR 16352.

7.10.10 Maintenance of the supported product by non-skilled and skilled persons

7.10.10.1 General

Separate information shall be provided for maintenance that can be performed by non-skilled and skilled persons, preferably as separate information products or topics.

7.10.10.2 Maintenance by non-skilled persons

If non-skilled persons are allowed to perform maintenance, the information for use shall include descriptions of those tasks with appropriate illustrations and an explanation of the nature and frequency of those tasks.

If some maintenance tasks need to be performed more frequently than others (e.g. every day) they shall be distinguished from periodic (short-term and long-term) maintenance and inspection activities.

Materials and tools used for cleaning, for example, appropriate chemical substances, cleaning cloths and brushes, shall also be specified. Possible risks arising from the use of inappropriate materials and tools shall be clearly specified.

When there are maintenance tasks that should not be attempted by non-skilled persons, the tasks shall be identified and information shall be given on how to contact the supplier or others to obtain technical assistance.

7.10.10.3 Maintenance by skilled persons

Information on maintenance by skilled persons shall include:

a) the nature and frequency of inspections;

b) safety precautions and warnings for maintenance carried out on running or live equipment;

c) drawings and diagrams enabling maintenance to be done effectively;

d) regular checking of warning devices;

e) details of cleaning methods;

f) required tools and materials;

g) safety hazards or problems resulting from corrosion or weakening of structural parts of the supported product;

h) schedules for planned periodic maintenance, or predicted maintenance based upon the results of the condition monitoring of the systems and components involved, or any combination of both methods; and

i) contact details of the supplier or others from whom technical assistance can be obtained.

7.10.11 Maintenance of complex systems

An overall maintenance and service procedure for a complex system such as an industrial plant should be drawn up by combining the maintenance and service procedures for key components.

Product maintenance includes procedures to prevent or correct errors. In systems, maintenance information includes the system as a whole as well as various subsystems and components, considering the required or expected system reliability and availability.

Maintenance information shall include the following:

a) safety precautions for performing maintenance such as complete or partial shutdown; and

b) procedures for accessing subsystems or components to perform maintenance, including means of emergency exit.

7.10.12 Troubleshooting and repair of the supported product by non-skilled and skilled persons

7.10.12.1 General

Separate information shall be provided for troubleshooting and repair that can be performed by non-skilled and skilled persons, preferably as separate information products or topics.

Contact details of the supplier or others from whom technical assistance can be obtained should be provided in a form that can be referred to without operation of the supported product.

7.10.12.2 Troubleshooting and repair by non-skilled persons

Where non-skilled persons are permitted to repair the supported product, the following information shall be provided:

a) method of repair without the risk of harm to persons or damage to the supported product; and

b) method for testing after the repair, if any test is required.

The information for use may include troubleshooting guidance or a list of frequently-asked-questions (FAQs), with appropriate illustrations, on the types of fault that non-skilled persons can reasonably be expected to attempt to repair.

7.10.12.3 Troubleshooting and repair by skilled persons

When repair of the supported product by skilled persons is permitted, the following information shall be provided:

1) method of repair; and

2) method for testing after the repair, if any test is required.

The following information shall be provided, where applicable:

a) list of indications for fault identification and its location;

b) list of indications of normal operation;

c) descriptions of built-in diagnostic systems to aid fault detection;

d) drawings and diagrams to enable troubleshooting to be carried out effectively;

e) directions for noting malfunctions of equipment, noting abnormal symptoms and indications, and recording operation of alarms and trips;

f) directions for starting standby or alternative systems, and for shutting down and isolating malfunctioning units;

g) repairs and adjustments;

h) aids for troubleshooting procedures, such as functionally identified maintenance systems, step-by-step fault diagnosis, algorithms, or, in complex systems, fault trees and computer-based fault diagnosis;

i) clear definition of troubleshooting and repair tasks that should not be attempted; and
j) contact details of the supplier or others from whom technical assistance can be obtained.

7.10.13 Replacement of parts of the supported product by non-skilled and skilled persons

7.10.13.1 General

Separate information shall be provided on replacement of parts that can be performed by non-skilled and skilled persons, preferably as separate information products or topics.

Contact details of the supplier or others from whom technical assistance can be obtained shall be provided.

7.10.13.2 Information on replacement of parts

Where replacement of parts is permitted, the following information shall be provided:

a) method of replacement without harm to persons or damage to the product;
b) method for testing after the replacement, if any test is required; and
c) source of spare parts.

If a refit of labels is necessary during the intended lifetime of the product, information about the locations of the labels and their replacement shall be given.

7.10.14 Disassembly, recycling, disposal of the supported product

7.10.14.1 General

Information for use shall provide information on disassembly, recycling or disposal of the product.

7.10.14.2 Disassembly

If disassembly by the target audience is permitted, information shall be provided, taking into account safety and environmental considerations.

7.10.14.3 Recycling

Where specific procedures are necessary for recycling of the product or its components, this information shall be provided in accordance with the appropriate legal requirements and standards.

7.10.14.4 Disposal

Information on disposal shall be provided, taking into account environmental and safety considerations, and legal requirements.

For products containing data or intellectual property such as software, the supplier shall identify in the information for use:

a) responsibilities concerning disposal;
b) determination of ownership of stored data and IT assets;
c) procedures to remove and retain or to destroy data and IT assets; and
d) procedures to confirm that data has been removed from data storage devices before they are recycled, or to physically destroy the devices.

7.11 Safety-related information

7.11.1 Types of safety-related information

Three types of safety-related information shall be separately identified:

a) safety signs and product safety labels;

b) safety notes; and

c) warning messages.

Safety-related restrictions concerning the use of the supported product shall be given in the information for use.

Safety-related information shall be incorporated into the information for the intended safe use of the product, including:

1) the intended environment;
2) type of application;
3) needed tools;
4) protective features of the product to be installed or activated;
5) safe use and disposal of the product;
6) working materials and consumables;
7) identification of permitted modifications of the product; and
8) procedures to maintain the product in a safe condition and to dispose of it safely.

7.11.2 Location of safety-related information

Safety notes shall be provided in a separate clause or section at the beginning of the information for use. This clause or section shall be identified and shall have a heading which emphasizes the importance of the content.

NOTE 1 It is not necessary and usually not helpful to place safety notes before the table of contents.

In step-by-step instructions, warning messages shall be given in the context in which a hazard is likely to occur.

If personal protective equipment is necessary, its information shall be described in the following places:

a) at the beginning of step-by-step instructions for a task that requires personal protective equipment; and

b) on the package and the product, where applicable.

NOTE 2 It is possible that product-related standards and legal regulations contain additional requirements.

7.11.3 Precautions for particular target audiences

Hazards for particularly vulnerable target audiences shall be identified at the beginning of the information for use. Such information shall be highlighted for emphasis. Vulnerable target audiences can include the following:

a) children;

b) older persons;

c) persons with disabilities;

d) persons with electrically controlled medical devices;

e) persons who suffer from allergic reactions to specific chemicals; and

f) persons who are sensitive to strobe lighting effects.

Appropriate warnings shall be provided where the use of the supported product has the potential to affect electrically controlled medical devices, such as heart pacemakers.

Advice shall be given to recognize adverse medical symptoms that are known to arise from the use of the supported product and to understand what preventive or remedial actions to implement.

7.11.4 Safety notes

Safety notes shall, as a minimum, indicate potential hazards and, if appropriate, their causes; the probable consequences of not avoiding them; and how to avoid such hazards.

Safety notes shall start with a general statement that highlights the importance of reading them.

EXAMPLE: To avoid possible harm, read and follow these instructions.

Safety notes shall include information covering:

a) identification of hazards with appropriate graphics, including:
    1) reasonably foreseeable misuse;
    2) inadvisable modifications of the product;
    3) when the product is no longer safe to use due to effects of wear or damage;
    4) radiation, including any sealed hazard, for example, sources of ionizing radiation, laser (in accordance with IEC 60825-1), microwave, ultraviolet, infrared and lethal voltages;
    5) substances which are carcinogenic, mutagenic or reprotoxic or other substances for which there is scientific evidence of probable hazard to human health; and
    6) emitted sound pressure levels;
b) harmful potential consequences of unsafe use;
c) safety precautions (means of avoiding hazards), such as protective clothing or goggles; restrictions on the personnel who can use the product, for example, by training, skills, or age; observance of product controls and messages; safety signs and labels; safekeeping of keys and passwords; and
d) means of recovery from hazards, including emergency procedures and use of emergency equipment.

7.11.5 Warning messages

7.11.5.1 General

Warning messages shall, as a minimum, concisely indicate potential hazards, and if appropriate, their causes; the probable consequences of not avoiding them; and how to avoid such hazards. Each warning message shall be placed either before the first step or incorporated just before the relevant step of a step-by-step instruction. If the information about the consequences of the hazards or the avoidance measures is readily understood by the target audience, this information may be omitted in the warning message itself. Warning messages on how to avoid hazards shall be concise. Detailed safety procedures should be presented elsewhere in the information for use as applicable.

NOTE Sometimes related hazards can be treated in a single warning message.

7.11.5.2 Design

The design of the warning messages shall be consistent and warning messages shall be made conspicuous and prominent. The following shall be taken into account:

a) start with signal words;
b) limit the text and illustrations to the essentials;
c) make the location, content and style of the warning messages conspicuous;
d) make warning messages visible to the target audiences and to any other people exposed to hazards, from their position during use, and at the right time; and
e) place warning messages as determined by the risk assessment.

7.11.5.3 Signal words for harm to persons

Signal words should be used.

EXAMPLE: Signal words are set out in ISO 3864-2, as follows:

DANGER: Signal word used to indicate an imminently hazardous situation which, if not avoided, will result in death or serious injury

WARNING: Signal word used to indicate a potentially hazardous situation which, if not avoided, could result in death or serious injury

CAUTION: Signal word used to indicate a potentially hazardous situation which, if not avoided, could result in minor or moderate injury

However, such distinctions can be too subtle to affect behaviour (or to be translated). In certain circumstances, signal phrases such as "DANGER OF DEATH", "RISK OF BLINDING", or "BEWARE OF FUMES" can be more effective at drawing greater attention to some instructions or safety information than signal words.

7.11.5.4 Signal words for damage to property

Potential risks of damage to the supported product (or to other property) should be indicated by other signal words such as "NOTICE", "ATTENTION", or "TAKE CARE" that are appropriate to the potential damage or to the precaution advised.

7.11.6 Safety-related information in quick-start guides

Quick-start guides shall include safety-critical information relevant to the tasks described in the quick-start guide. These shall also make clear to the target audiences why and when they need to read additional safety-related information in the information for use.

Quick-start guides shall include reference to the detailed safety-related information in the instructions. The reference shall be unambiguous and presented conspicuously at the beginning of the quick-start guide together with the general warning sign (ISO 7010-W001:2011-05) or a signal word. The text of the reference shall describe the possible consequences of not complying with the general safety messages.

7.12 Instructions for assembly of self-assembly products

The instructions for assembly of self-assembly products shall present the sequence of assembly actions in the form of one-step-at-a-time sequence of illustrations, each accompanied by text positioned adjacent to that step. Where necessary, explanatory or advisory information and warnings shall be provided. Additional supporting assembly guidance may be provided in other media, for example video demonstrations. The information for use shall specify instructions for each assembly step as follows:

a) the components and other objects involved in that step;
b) how components are to be aligned and oriented;
c) the actions the assembler needs to perform to shape, attach or connect the components;
d) the appropriate tools required;
e) safety precautions to take or foreseeable errors to avoid during this step; and

f) how to confirm correct mating of components and that their attachment has the required strength.

Consumers find it easier to understand assembly actions illustrated in 3-dimensional line drawings than in photographs or 2-dimensional views. The directions in which components need to be aligned or moved towards mating points should usually be indicated by arrows.

If assembly or erection requires special skills, this shall be clearly indicated in the information for use by a warning such as "for assembly by a qualified electrician" or "assembly by adults required".

Foreseeable errors by an assembler that probably compromise the safety of the completed product – or the safety of a person while undertaking the assembly – shall be subject to risk management (see 6.2.7). Assessment of whether the instructions adequately address these risks should include usability testing. See also 6.3.3.

7.13 Information for use for a complex system

The configuration of the relevant components of a complex system provides new functions and conditions to be addressed beyond the information for use of each subsystem and component.

For complex systems, each system as a whole shall include information for use, including restrictions for use and safety notes on the overall system level. For operational tasks, the step-by-step instructions for a system shall be structured to present the users' tasks, not only to describe the relationship of the system components.

NOTE IEC 81346-1 presents one way of structuring descriptions of complex systems, according to descriptions of the subsystems and components.

7.14 Information security and data privacy

7.14.1 General

Where products contain data, the supplier shall include instructions for features intended to protect the confidentiality, integrity, and availability of data from loss or misuse. For remotely controlled or network-connected products, information for use shall include features to avoid misuse or unauthorized access (hacking), such as changing default passwords.

EXAMPLE User controlled product settings for protection or exposure of personal data.

7.14.2 Functions for access control or protection of sensitive data

The most common methods of providing information protection or assurance shall be taken into account when creating, exchanging, delivering or storing information.

These are:

a) Encryption: a software-based security solution for protecting data from unauthorized access or theft. However, the information can be corrupted by malicious programs or an intruder making the information unrecoverable. Hardware-based security solutions can prevent read and write access to data and hence offer protection.
b) Backups: used to retrieve data which is lost from another source. It is therefore essential to have backup systems in place.
c) Data masking: the process of obscuring (masking) specific data within a database table or cell to ensure that information security is maintained and sensitive information is not exposed to unauthorized personnel.
d) Data erasure: a method of overwriting information that completely destroys all electronic data residing on a hard drive or other digital media to ensure that no sensitive information is leaked when an asset is retired or reused.

ISO/IEC 27002 should be taken into account when considering the issue of data security in information.

7.15 Training

Where training is required for the target audiences, this shall be specified in the information for use.

8 Structure of information for use

8.1 General

Information for use shall be structured so as to enhance its usability and comprehensibility, with, as applicable, features allowing for easy search, convenient navigation, and unambiguous understanding of its contents.

Where general information for use is likely to be lengthy or complex, it shall be clearly divided into convenient parts, and shall have a consistent format.

Information for use should be separated into chapters or sections if there are distinctly different target audiences, for example, installers and repair technicians.

For printed information for use comprising multiple documents, the information on the front covers or spines shall make it easy to distinguish one document from another.

NOTE Further information on structuring and documentation can be found in IEC 81346-1 and IEC 62023.

Where supplementary information is provided, it may be indicated on a prominent place of the information for use.

If supplementary information is available, the information for use can provide a location at which it can be accessed, for example a link to a website containing supplementary information.

A system can consist of different elements from various suppliers. If a supplier of such a system develops the information for use for the system, the supplier may integrate the provided information for use of the elements as a part of the overall information for use.

The following shall be considered when information from multiple sources is integrated:

a) consistent document numbering;
b) methods of integration of information for use, for example, integration of a copy, adaptation of the content, cross-reference to the source document, or a specific section in the source document; and
c) keeping terminology consistent or using cross-references.

8.2 Information types

Information for use should be allocated in its functional structure to predefined information types, such as:

a) conceptual information. This information type includes the following:
    1) information about the information for use itself;
    2) description and identification of product and components;
    3) description of controls and displays;
    4) process description (use case or concept of operation); and

   5) safety note.

b) instructional information. This information type includes the following:
   1) step-by-step procedure (e.g. for assembly, installation, maintenance tasks); and
   2) warning message (as part of the procedures).

c) reference information. This information type includes the following:
   1) title page information;
   2) legally required information;
   3) troubleshooting;
   4) maintenance schedule;
   5) table of contents;
   6) glossary of terminology; and
   7) index.

8.3 Structuring

8.3.1 General

Structuring means allocating information to an information model or framework in accordance with the principles and detailed requirements in this document. Information for use shall be structured on all levels of information:

a) a set of information products if there is more than one information product;
b) the information product; and
c) the information type.

To help to ensure consistent structure, rules should be defined and applied to the information for use.

NOTE At an additional level, structuring can be applied to elements, such as sentences, figures, media objects, and mark-ups of specific object types.

Appropriate rules for the structure shall be applied on each level as far as applicable.

8.3.2 Use of information models

There are three approaches for the structuring of content listed below, one of which should be applied:

a) developing an information model by using a structuring method;
b) using an existing information model, for example an open source information model; and
c) adapting an existing information model by using a structuring method.

8.3.3 Use of leading criteria

Selection of leading criteria for structuring should depend on factors such as the supported product, the information product and the target audiences. Examples for leading criteria are listed in Table 2:

Table 2 – Structuring principles

Principle	Consideration for subdividing and arranging topics
Task	in the order that tasks are performed
Product	the product's functions or elements
Product life cycle	stages, e.g., shipping, installing, operating, servicing, repairing, disposing
Target audience	segmentation of information according to audiences' information needs
Cognitive	important first, from simple to complex
Parameter	alphabetic sort for ease of reference, e.g., parameter lists, indexes

8.3.4 Detailed structure of step-by-step instructions

8.3.4.1 General

Step-by-step instructions shall include preliminary information, instructional steps, and completion information.

8.3.4.2 Preliminary information

Preliminary information shall include the following:

a) a brief overview of the purpose of the procedure and definitions or explanations of necessary concepts not included elsewhere;

b) identification of technical or administrative activities required before starting the task;

c) a list of resources the target audience needs in order to complete the task, which can include tools, other people, data, documents, passwords, or additional software or equipment; and

d) relevant warnings, cautions, and notes that apply to the entire procedure.

Preliminary information common to several steps can be grouped and presented once to avoid redundancy.

8.3.4.3 Instructional steps

Instructional steps shall be numbered using Arabic numerals and presented in the order of performance.

Relevant warning messages and cautions shall be provided before the applicable instructional steps.

Each step provides a single action. Instructional steps shall indicate the expected result or system response and explain error messages and recovery procedures which could occur.

Lengthy or complex step-by-step instructions should be segmented. Alternative or repeated steps should be clearly indicated, to enable the target audiences to determine which alternate or repeated steps to perform or skip and where to rejoin the main procedure.

Notes should indicate steps where reference information should be consulted (e.g. special codes or commands), or steps requiring special attention or where errors are common.

In English, French, and other languages where the imperative exists and is culturally acceptable, the instructional steps should use the imperative form to describe the target audience's action.

8.3.4.4 Completion information

Information for use shall indicate the end of a set of step-by-step instructions. It should explain how the target audience determines that the procedure has been successfully completed (this is often a consideration in product design and testing), and may include references or links to related topics, such as subsequent tasks or troubleshooting.

8.4 Navigation and information delivery

8.4.1 General

Information for use shall include elements that make it easy to navigate the contents, as specified in the remainder of 8.4.

8.4.2 Navigating printed information for use

8.4.2.1 Page numbering

Where information for use comprises more than two (printed) pages, the pages shall be numbered. A numbering as (n) of (m) where (n) indicates the actual page number and (m) the total number of pages may be used.

8.4.2.2 Table of contents

Information for use that exceeds 12 pages shall have a table of contents unless it can be demonstrated not to be necessary. Headings and page numbers appearing in the table of contents shall be the same as those used in the text. The table of contents should have enough levels to help the target audience to find information efficiently but not exceed four levels of headings.

8.4.2.3 Index

Where information for use is lengthy and complex, an index of keywords presented in alphabetical order should be included. The index shall be referenced in the table of contents.

NOTE More information about indexing can be found in ISO 5963.

8.4.3 Dynamic delivery

8.4.3.1 Individualized delivery

If possible, only the particular piece of information needed to attain a specified goal or perform a specific action should be displayed to the target audiences. Additionally, full information regarding a supported product should be delivered to the target audience by a separate method.

If topics in the information for use are restricted to specific target audiences (e.g., paid customer, user group or user profile), the information delivery system should limit access to the allowed target audience.

8.4.3.2 Context sensitivity

If context information (e.g. geographical data, voice data or visual patterns) is accessible by the individual's electronic device, the information for use may use the context information to deliver more precise information to the target audience, subject to their consent.

8.4.3.3 Search functions

A mechanism for electronic search shall be provided in information for use delivered electronically. A search function shall at a minimum include text strings. The search function should not only be based on full text search, but also be supported by search algorithms, taxonomies, thesauruses, and other available technologies.

8.4.3.4 Links to related topics

If appropriate, each topic of electronic information for use may be linked to other relevant topics, but any linking shall not distract or demotivate the target audiences. Different electronic media (e.g. text, video, or auditory elements) dealing with the same or similar subjects, shall be linked or cross-referenced so that they can be used together if appropriate.

External links (URLs) are considered volatile and should not be used as the only way of referencing, although they may be supplied as an additional service.

In printed information for use, the complete URL for related information should be included for ease of reference.

9 Media and format of information for use

9.1 General

The supplier shall set editorial style conventions for the format of information for use and for the media to be used. The supplier shall determine the media and format of the information for use according to the nature of the target audiences and based on their needs. The media and format shall allow easy access to information for use throughout the intended lifetime of the product.

The following are examples of possible media for information for use:

- text, photographs, safety signs, graphical symbols and illustrations;
- video (including auxiliary means such as audio and subtitles);
- animated illustrations and 3D-models controllable by the user;
- speech;
- braille, tactile figures, signs or other visible hand gestures;
- augmented reality (information embedded in an electronic representation of the current environment);
- virtual reality;
- combinations of the above;
- leaflets (e.g. paper, plastic), stapled booklets or bound manuals with text and illustrations; and
- printed information for use on the packaging or on the product itself (as permanent marking or peelable tag); and printable versions of documents which are made readable on-screen through a file exchange format in which the layout matches that of the printed version.

When any part of the information for use is made available on the Internet, both the presentation and the necessary methods of navigation should conform to ISO/IEC 40500:2012.

9.2 Durability of chosen media

Information for use printed onto or moulded into the supported product shall remain legible throughout the intended lifetime of the supported product.

Information for use given only on packaging or in materials accompanying the supported product shall be in durable form and be able to survive frequent use during the intended lifetime of the supported product and in the environment where the supported product is intended to be used.

9.3 Use of animation or audio-visual demonstrations

Duration, information density or complexity of electronic sequences shall be adapted for the target audience.

Information for use in electronic media form shall follow the order of activities that the target audience has to perform. Animation sequences shall be accompanied by synchronized spoken or written text, or both.

Advertising or promotional content shall be kept separate from the information for use and viewing of such content shall not be required prior to viewing the information for use.

Where complex processes need to be explained, a sequence of visual and auditory elements can help to understand the nature of the task.

9.4 Location and availability

Information for use shall be given in one or more of the following locations as appropriate:

a) inside the packaging of the supported product;

b) on or within the supported product;

c) on the packaging – but not only on the packaging;

d) on websites provided by suppliers or others; and

e) as separately provided collateral documentation.

Information concerning installation, troubleshooting, repair, performing recovery actions, and contacting technical support shall be provided in a form that is available when the supported product is not operating.

If the information for use is complex, important messages should be displayed on the supported product in addition to the locations set out above, for example, by means of short reference or reminder cards, stickers, illustrations, or labels.

9.5 Printable information

Page design applies to printed documents and electronic documents which can be printed.

Page design should include consideration of the following:

a) legibility when printed on common paper sizes (e.g. A3, A4, A5);

b) number of pages required to deliver complete information; and

c) number and size of photographs, safety signs, graphical symbols, diagrams, and tables.

9.6 User interaction and search features

Electronic media applications shall take advantage of dynamic navigation and presentation devices to facilitate rapid location, comprehension, and, where appropriate, printing of the required information.

Electronic media applications shall provide navigation systems (preferably tree-like) that facilitate free movement within an application, as well as access to individual sections of the application. Menus, buttons, hyperlinks, breadcrumb trails, and bookmarks shall be provided to enable orientation and interaction with the system.

NOTE A breadcrumb trail is a navigational aid with a displayed series of hyperlinks which lead from the home page to the current page, allowing the user to return to previously viewed pages.

Electronic information shall be searchable. The search features should allow common synonyms to be found. The search feature should also show results in the case of common misspellings.

9.7 Downloadable information for use

If information for use is intended to be downloaded in electronic form, it shall be provided in a form that allows for display on commonly used devices and operating systems. In this case, websites from which information for use can be downloaded shall:

a) provide either the software needed, or a link for downloading the required viewing software; and
b) allow downloading at any time.

NOTE More information can be found in ISO/IEC/IEEE 23026.

9.8 Suitability for the conditions of use of the supported product

The conditions of use of the supported product shall be addressed when choosing the media and format for information products. Table 3 lists examples of considerations for the choice of media, depending on the conditions of use of the supported product.

In case of doubt, the adequacy of media shall be tested using the information for use in a realistic scenario.

Table 3 – Examples of considerations for the choice of media

Conditions of use of the supported product	Inadequate media for the information for use
Low-light conditions	Text without light (paper, electronic ink)
Noisy environment Ear protection	Speech
Wetness, humidity Cleanroom conditions	Paper

9.9 Consistency of format

Consistent format shall be used throughout the information for use, for the user interface, data elements, field names, functions, pages, topics, and processes.

Agreed formatting conventions shall be applied consistently throughout the information for use. These include colours, borders, indenting, spacing, and font variations.

9.10 Legibility, readability and comprehensibility

9.10.1 Text font sizes and heights of safety signs and graphical symbols

Text fonts, safety signs and graphical symbols shall be clearly legible for the target audiences, whether printed, integrated with the supported product, or computerized.

Recommended minimum text font sizes and heights of safety signs and graphical symbols are listed in Table 4. The text font size shown in Table 4 are applicable for printed material, including product labels. Where electronic displays cannot be enlarged, equivalent text sizes should be used.

NOTE The best legibility is achieved when optimizing the relationship between variables, including type font, font size, contrast, number of characters per line (approximately 70 in Latin characters) and line spacing (minimum 120 % of type size in Latin characters).

Where measures such as bold lettering, underlining, or upper case and italic lettering are used for emphasis, these measures shall be used carefully and sparingly so as not to reduce legibility.

Suppliers shall make provisions for a target audience with sight or hearing impairments to access information for use.

Table 4 – Minimum recommended text font sizes and heights of safety signs and graphical symbols

NOTE The term "point" (abbreviated here to "pt") is a unit of measurement of type sizes and spacing. There are several kinds of standards. However, "DTP point" is adopted in this document which is 0,353 mm or 1/72 inch.

Product/ information document size	Location and role of instruction	High contrast dark text on light background	Low contrast colours or white on black	Complex character sets (e.g. Kanji)	Other remarks	Safety signs and graphical symbols: Symbols generally	Safety signs and graphical symbols: Safety signs
Instructions viewed from up to 1 m distance on floor-standing products	Critical on-product markings	14 pt bold **BE68.3 1,0QGO aeocld**	16 pt bold **BE68.3 1,0QGO aeocld**		Consider using large print fonts specially developed to help people with visual impairments to read signs and labels at 30 cm to 100 cm	As required by standards/regulations Otherwise according to viewing distance from which attention needs to be attracted or the symbol needs to be recognized. Less than 15 mm height unlikely to be sufficient for critical on-product markings	
	text	14 pt	16 pt				
Instructions on desk-top products, information in printed manuals or single-fold leaflets and documents intended for printing by user	Critical on-product markings	14 pt bold **BE68.3 1,0QGO aeocld**	16 pt bold **BE68.3 1,0QGO aeocld**				
	Headings	12 pt BEFPR 68.391,0 QGOC aeocld	14 pt BEFPR 68.391,0 QGOC- aeocld		Serif fonts may be used (but sans-serif preferred)	5 mm min. height (or 14 pt): e.g. ☺	10 mm min. height: e.g. 
	continuous text	10 pt BEF68.39 aeocld	12 pt BEF68.39 aeocld			Do not use safety signs or graphical symbols in < 14 pt continuous text	
Hand-held products and multi-fold instruction sheets	Critical on-product markings	12 pt BEF68.39 1,0QGOC aeocld	12 pt bold **BEF68.39 1,0QGOC aeocld**	9 pt with 150 % line spacing 電気 規格		5 mm min. height	10 mm min. height
	Headings, decimals	10 pt BEFPR 68.391,0 QGOC aeocld	12 pt BEFPR 68.391,0 QGOC aeocld		Use only sans-serif fonts at this level and below	Pref. min. 5 mm; 4 mm/12 pt if very simple	10 mm min. height[a]
	continuous text	9 pt BEFPR 68.391,0 QGOC aeocld	Not advised for text smaller than 12 pt		**Electronic, audio or LARGE-PRINT media should be available on demand** (e.g. from website or point of sale)	Do not use safety signs or graphical symbols in < 14 pt continuous text	
Very small products and packaging (e.g. with printable surface $< 10\ cm^2$)	Markings, headings, warning messages, decimals	8 pt BEFPR 68.391,0 QGOC aeocld		8 pt with 150 % line spacing 電気 規格		Pref. min. 5 mm; or 3 mm if very simple ☺	10 mm min. height[a]
	continuous text	6 pt BEFPR 68.391,0 QGOC aeocld				Do not use any symbols in < 10 pt continuous text	

[a] except for the general warning sign (ISO 7010-W001:2011-05) accompanying a text warning, which may be: 5 mm minimum as a marking ⚠ and 3 mm minimum in headings to text. ⚠

9.10.2 Maximum brightness contrast

For printed, embossed or engraved information for use, the brightness contrast (ratio of reflectance of the bright part over reflectance of the dark part) shall be as high as possible.

Contrast requirements for text displayed on visual display terminals shall be in accordance with ISO 9241-300.

9.10.3 Information for use provided on the supported product or packaging

The location of on-product information for use and the angle between its surface and the vertical plane shall be such that it is easily legible from the intended position of the user or initial encounter. Where a supported product is intended for a target audience with particular needs, this should be taken into account when determining the location of information for use.

Information for use on packaging should be visible and legible when the package is in the upright position.

The advantages in terms of durability and production of incorporating information for use in the material (i.e. moulded-on) of the supported product should be weighed against the disadvantages of loss of contrast and legibility.

NOTE Incorporating information for use in the material of the supported product generally produces an inferior result compared with that of good quality printing.

9.10.4 Minimum heights of safety signs and graphical symbols

Minimum heights for safety signs and graphical symbols are listed in Table 4. The heights of safety signs and graphical symbols should not be smaller than these heights unless smaller heights are unavoidable because of the dimensions of the supported product. Smaller heights should only be adopted after careful consideration of possible safety consequences for the target audience.

9.10.5 Rules for simple wording

Information for use should be written in simple, clear, direct phrases within short sentences on the principle of "one expression, one simple meaning". Colloquial expressions and uncommon regional variations of names and product features should be avoided.

9.10.6 Function of information sections

The function of information sections (for example, the description of the product or information about troubleshooting) should be readily identifiable. Different kinds of information should be clearly identified.

9.11 Use of visualization

9.11.1 Graphical symbols and safety signs

Where applicable, graphical symbols in diagrams shall conform to IEC 60617 and to ISO 14617 (all parts). Symbols in flowcharts shall conform to ISO 5807. If a graphical concept is represented by a graphical symbol registered in ISO 7000 or IEC 60417, the registered graphical symbol shall be used.

NOTE Rules for drawing-specific diagrams are given in IEC 61082-1, ISO 5807, and ISO 15519-1.

Graphical symbols and safety signs provide a language-independent means for conveying safety information. Only registered safety signs in ISO 7010 shall be used in information for use.

Safety signs shall comply with the design principles (including shapes and colours) specified in ISO 3864 (all parts).

Safety colours specified in ISO 3864 (all parts) shall be used only for safety signs.

9.11.2 Illustrations

Where a sequence of operations is described, the text and illustrations shall follow the operational sequence. Illustrations shall be placed as close as possible to the text to which they relate, so that they can be viewed adjacent to the relevant text.

Where illustrations need explanatory texts, the text shall be placed adjacent to the illustration. A sequence of illustrations should be logical and comprehensible. Related text and illustrations should both be viewable at the same time, each supporting the other to enhance comprehensibility.

Illustrations should be suitable for the target audiences in the target cultures. Illustrations should be composed so as to focus attention on important details and to be self-explanatory. Illustrations should be chosen or designed individually for each purpose by a competent graphic artist or technical illustrator. The print quality or screen resolution of illustrations should encourage rapid recognition. Resolutions below 72 dpi can significantly impede clarity. Resolutions above 300 dpi are preferable.

9.11.3 Information content of illustrations

Information overload in illustrations shall be avoided. Illustrations should normally provide only information relevant to describing the related function. Illustrations or detailed part(s) of these should be repeated in the relevant part(s) of information for use as needed.

9.11.4 Illustrations with captions

Illustrations should be identified and supplemented with captions. There shall be a clear relationship between the illustration(s) and the captions. Related illustrations and captions should be located in similar positions (e.g., illustration over caption, or caption over illustration, not mixed). Captions should be numbered for clear reference from the main text.

9.12 Use of tables

Tables shall be set out clearly, informatively, and in a consistent design. Tables shall be positioned next to the relevant text, except that extensive reference tables may be placed in annexes.

9.13 Use of colours

If the target audience includes people with colour-deficient vision, colours should be used only as a supplementary means of drawing attention to particular sections of information.

Information shown in contrasting colours should remain distinct if the target audience prints it on a monochrome printer.

9.14 Use of icons

Icons should uniquely represent objects or functions. Different icons should never be used for the same object or function. Similarly, different objects or functions should not be represented by the same icon. If an icon represents changes in the status of an item, the metaphor should be used consistently.

NOTE IEC 62744 provides information on how to use icons to present information on the behaviour of a supported product.

9.15 Attracting attention to safety-related information

9.15.1 General

Safety-related information shall be emphasized by means such as an initial signal word (see 7.11.5.3 and 7.11.5.4) in larger or different type font or size, use of colours, safety signs or graphical symbols, lines, indents, or other means of making it conspicuous.

Warning messages about hazards or restrictions on use are crucial for safety and shall be displayed at least as prominently as the other information in documents issued with the supported product.

EXAMPLE: **NOT suitable for children under three years of age.**

Safety notes shall be emphasized by a headline that identifies the subsequent material as safety-related. Safety notes shall be placed in the information for use in a prominent position before the use of the supported product is addressed in the information for use. In printed information for use, safety notes shall be presented in the front part and before information for transport, assembly, operation, etc., is given.

If warning messages are integrated between the steps of a procedure, the formatting (especially excessive highlighting) of the warning message should not distract the target audience from reading the information.

Warning messages displayed on a screen shall be visible for the whole time where they are relevant. Displayed warning messages may require user acknowledgement before they are removed from view.

Illustrations intended to convey safety-related information shall use colours, shapes, positioning or other means to make the information conspicuous. Where appropriate, safety signs and graphical symbols should be used to emphasize key messages (see 9.11.1).

9.15.2 Durability and visibility

Warning messages shall be visible for the intended lifetime of the supported product. When colours are used in warning messages, avoidance of discoloration and fading shall be addressed during the preparation of information for use.

10 Professional competencies

10.1 General

The creation of information for use shall be assigned to competent persons. Organizations should analyse their content creation process, identify the tasks to be performed to achieve the required results, evaluate the competencies which are needed to perform these tasks successfully, and designate the tasks and responsibilities to persons who cover these competencies. Responsibilities should be assigned to separate persons where the size of the organization permits.

The extent to which tasks and roles can be differentiated depends on the resources available to the organization. In some cases, only one person may be assigned to cover all tasks and take most of the roles and responsibilities. In larger organizations, teams or departments for the creation of information for use can be established.

10.2 Task-related competencies

To maximize the comprehensibility and safe use of a product, tasks related to information for use shall be assigned to competent persons. The creation of information for use should involve the following task-related competencies:

a) conduct research and analyse the life cycle of the supported product, the target audiences (the users of the product), the information that has to be given, and the language(s) in which the information will be used;

b) develop standard processes and methods for developing information products efficiently, effectively and in a consistent way;

c) develop a content strategy and a concept for the media for delivery of the information for use, defining appropriate content structures templates, and styles; check that existing information products are valid and up to date;

d) implement necessary enabling systems, for example for content management and information management;

e) gather safety-related information from the developers of the supported product;

f) develop written, audio, or visual content, including safety-related information;

g) perform proofreading and checking of the information for use;

h) use the appropriate methodologies and tools to make the information management process more efficient and effective;

i) organize translation and localization of the information for use;

j) approve and test information for use according to formal and functional quality and usability standards and criteria, and validate the correctness of the content;

k) produce output media or instructional applications and make the information for use ready for delivery;

l) analyse the way in which the target audience benefits from the information for use by appropriate methodologies and tools such as website statistics, empirical tests and social media feedback; and

m) provide leadership, make project plans, manage service providers, and communicate with developers and customer-facing associates within the supplier's organization.

The competent persons responsible for the information for use should be assigned to designated roles within the information creation process.

10.3 Level of proficiency

10.3.1 General

Various levels of proficiency should be identified. The following subclauses show a typical scheme for proficiency levels.

10.3.2 Proficiency level 1

Responsibilities of proficiency level 1 should include the use and application of relevant standards to define the information structure and develop or update information. This involves:

a) a wide range of theoretical and factual knowledge in the area of work;

b) a set of cognitive and practical skills to find solutions to specific problems;

c) the ability to accept responsibility for accomplishing work; and

d) adapting one's own behaviour to suit the relevant circumstances when problem-solving.

10.3.3 Proficiency level 2

Responsibilities of proficiency level 2 should include the ability to determine the requirements for the information products, taking into account the purpose and domain to which they apply. This involves:

a) extensive, specialist theoretical and factual knowledge in the area of work and an awareness of the limits of this knowledge;

b) practical abilities, such as extensive cognitive and practical skills to devise creative solutions to abstract problems;

c) the ability to act autonomously within the guidelines of work contexts that, as a rule, are generally known but nevertheless liable to change; and

d) overseeing the routine work of other persons, accepting a certain amount of responsibility for assessing and improving their work.

10.3.4 Proficiency level 3

Responsibilities of proficiency level 3 should include the ability to manage the process of information creation, develop content strategies, take responsibility for conformity to applicable standards and policies, and lead information development teams. This involves:

a) advanced knowledge of a field of work, involving critical understanding of theory and principles;

b) practical abilities such as advanced skills, demonstrating mastery and innovation, needed to solve complex and unpredictable problems in the specialized field of work; and

c) the ability to manage complex technical or professional activities or projects, taking responsibility for decision-making in unpredictable work contexts; and for managing professional development of individuals and groups.

10.4 Competencies of translators

When information for use is translated from the original language into a target language, competent translators or language specialists shall be responsible for the translation, including checking and proofreading.

Translators should have basic competencies as in 10.3.2 and should be fluent in both the original and the target language. They should be native speakers in the target language and should be familiar with the type of product and any product-specific terminology.

Annex A
(informative)

Guidance on evaluation

A.1 General

Annex A provides guidance on methods for evaluating the fulfilment of requirements of this document. As stated in 4.1, the requirements of this document are of two kinds, which may be assessed using the following methods:

a) fulfilment of requirements with regards to the information for use:
   1) comprehensiveness check;
   2) inspection for effectiveness (desk check); and
   3) empirical effectiveness check;
b) fulfilment of requirements with regards to the information management process:
   1) process assessment; and
   2) competency assessment.

NOTE A legal check and checks for compliance with internal documentation and product-specific standards can also be useful. Legal requirements for the retention of internal product documentation in the jurisdiction where the supported product is marketed can be researched and assessed. This can result in a traceable list of requirements and allow the information for use to be assessed on the basis of those requirements.

A.2 Assessing fulfilment of requirements for information for use supporting a particular product

A.2.1 Comprehensiveness check

Fulfilment of the requirements in Clause 7 should be assessed by a suitably qualified and experienced person conducting a comprehensiveness check to assess whether or not the required content has been included in the information for use. If any requirements in Clause 7 are not applicable to the supported product, the evaluation record should so state, with reasons, for example, the requirement that the weight of a product has to be specified is not applicable for downloaded software. For such requirements, the result may be N/A (not applicable).

A.2.2 Inspection for effectiveness (desk check)

Fulfilment of the requirements in Clauses 8 and 9 for structure, media and format of information for use should be evaluated by a suitably qualified and experienced person. Where fulfilment of a specific requirement is incomplete, but overall the information for use is judged as fulfiling Clauses 8 and 9, the reasons should be explained in the record, referring to the principles in Clause 5.

A.2.3 Empirical effectiveness check

Empirical evidence should be used to demonstrate that the information for use is effective for the target audience. For examples of empirical methods, see Table 1.

Empirical effectiveness checks should be designed as usability tests of the information for use, together with the supported product. Where the complexity of a product creates difficulties in performing usability tests, empirical effectiveness checks may be performed by experts following and understanding the information for use together with the supported product.

A.2.4 Useful additional checks

A.2.4.1 Residual risk check

The information for use should be evaluated to determine if residual risks from the product risk assessment have been addressed. For example, it should be checked whether the information for use include appropriate warnings for hazards.

A.2.4.2 Product-specific standards check

Information for use should conform to the relevant product-specific safety standard(s).

The relevant product-specific safety standards should be researched and evaluated in advance. A traceable list of requirements should be created; the information for use then should be assessed on the basis of the list of requirements.

Records of the research and evaluation of relevant product-specific standards against which the information for use was compared should be prepared and retained.

A.3 Evaluating the fulfilment of requirements for an information management process

A.3.1 Process evaluation

Fulfilment of the requirements in Clause 6 should be evaluated by a suitably qualified and experienced person. Records of the evaluation should provide evidence that the process of information management meets the requirements of this document.

The process evaluation should be repeated when the process is changed.

Methods of evaluation can include questionnaires, interviews with managers and other responsible persons, access to artefacts from the relevant information systems, and inspection of the internal documentation.

NOTE ISO/IEC 33001 provides requirements for process assessment.

A.3.2 Competency evaluation

The competencies of the technical writers responsible for the development of information for use, including checking and proofreading, should be evaluated. It is reasonable to perform the competency evaluation together with the process evaluation. The evaluation should consider interviews, documented evidence showing education or experience, continuing training certificates, and a training plan showing continuing training of technical writers on relevant topics.

A.4 Guidance on conducting an evaluation

A.4.1 Result of the evaluation and corrective actions

The results of each part of the evaluation should be documented, together with overall judgement as to fulfilment of requirements and a listing of corrective actions considered necessary, classified as follows:

1) specific findings of not or partially fulfilled requirements; and
2) recommendations to improve the information for use and the processes.

The evaluation should close with a clear statement of the suitably qualified and experienced person whether a claim of fulfilment of requirements may be made or not, together with the scope of this claim (information for use, validity period, and organization).

A.4.2 Evaluation of similar information for use (conferrable evaluations)

Information for use should be evaluated. Information products which are similar in layout to already successfully evaluated information products may be assumed to fulfil the requirements of this document if:

a) an adequate evaluation is performed for every part of the information for use of each supported product, for correctness and comprehensiveness;

b) the processes established by the organization to prepare information for use have been evaluated as satisfactory;

c) the layout (of pages and text) is secured by an automated layout process; and

d) the layout of an adequate sample of similar information products has been evaluated.

It is not necessary to evaluate the layout of every single information product if the layout is secured by an automated layout process. In this case the layout of an adequate sample of similar information products should be evaluated.

Bibliography

[1] IEC 60050-151:2001, *International Electrotechnical Vocabulary – Part 151: Electrical and magnetic devices* (available at http://www.electropedia.org)

[2] IEC 60050-651:2014, *International Electrotechnical Vocabulary – Part 651: Live working* (available at http://www.electropedia.org)

[3] IEC 60073, *Basic and safety principles for man-machine interface, marking and identification – Coding principles for indicators and actuators*

[4] IEC 60204-1, *Safety of machinery – Electrical equipment of machines – Part 1: General requirements*

[5] IEC 60335 (all parts), *Household and similar electrical appliances – Safety*

[6] IEC 60529, *Degrees of protection provided by enclosures (IP Code)*

[7] IEC 60848, *GRAFCET specification language for sequential function charts*

[8] IEC 61082-1:2014, *Preparation of documents used in electrotechnology – Part 1: Rules*

[9] IEC 61310-1, *Safety of machinery – Indication, marking and actuation – Part 1: Requirements for visual, acoustic and tactile signals*

[10] IEC 61355-1:2008, *Classification and designation of documents for plants, systems and equipment – Part 1: Rules and classification tables*

[11] IEC 61506, *Industrial-process measurement and control – Documentation of application software*

[12] IEC 62023, *Structuring of technical information and documentation*

[13] IEC 62507-1, *Identification systems enabling unambiguous information interchange – Requirements – Part 1: Principles and methods*

[14] IEC 62569-1, *Generic specification of information on products by properties - Part 1: Principles and methods*

[15] IEC 62744, *Representation of states of objects by graphical symbols*

[16] IEC 80416-1:2008, *Basic principles for graphical symbols for use on equipment – Part 1: Creation of graphical symbols for registration*

[17] IEC 81346-1:2009, *Industrial systems, installations and equipment and industrial products – Structuring principles and reference designations – Part 1: Basic rules*

[18] ISO 3864-2:2016, *Graphical symbols — Safety colours and safety signs — Part 2: Design principles for product safety labels*

[19] ISO 10377, *Consumer product safety – Guidelines for suppliers*

[20] ISO 10628-1:2014, *Diagrams for the chemical and petrochemical industry – Part 1: Specification of diagrams*

[21] ISO 10628-2:2012, *Diagrams for the chemical and petrochemical industry – Part 2: Graphical symbols*

[22] ISO 11429, *Ergonomics – System of auditory and visual danger and information signals*

[23] ISO 12100, *Safety of machinery – General principles for design – Risk assessment and risk reduction*

[24] ISO 14971, *Medical devices – Application of risk management to medical devices*

[25] ISO 15006, *Road vehicles – Ergonomic aspects of transport information and control systems – Specifications for in-vehicle auditory presentation*

[26] ISO 15519-1, *Specification for diagrams for process industry – Part 1: General rules*

[27] ISO 17100, *Translation services – Requirements for translation services*

[28] ISO 17842-1:2015, *Safety of amusement rides and amusement devices – Part 1: Design and manufacture*

[29] ISO/DIS 20607:2018, *Safety of machinery – Instruction handbook – General drafting principles* [1]

[30] ISO 31000, *Risk management – Guidelines*

[31] ISO 5963, *Documentation – Methods for examining documents, determining their subjects, and selecting indexing terms*

[32] ISO 639-2, *Codes for the representation of names of languages – Part 2: Alpha-3 code*

[33] ISO 704, *Terminology work – Principles and methods*

[34] ISO 7731, *Ergonomics – Danger signals for public and work areas – Auditory danger signals*

[35] ISO 9000, *Quality management systems – Fundamentals and vocabulary*

[36] ISO 9186 (all parts), *Graphical symbols – Test methods*

[37] ISO 9241-210:2010, *Ergonomics of human-system interaction – Part 210: Human-centred design for interactive systems*

[38] ISO Guide 73, *Risk management – Vocabulary*

[39] ISO 9241-940:2017, *Ergonomics of human-system interaction – Part 940: Evaluation of tactile and haptic interactions*

[40] ISO/IEC 25041:2012, *Systems and software engineering – Systems and software Quality Requirements and Evaluation (SQuaRE) – Evaluation guide for developers, acquirers and independent evaluators*

---

1 Under preparation. Stage at the time of publication: ISO/DIS 20607:2018.

[41] ISO/IEC 26514, *Systems and software engineering – Requirements for designers and developers of user documentation*

[42] ISO/IEC 27002, *Information technology – Security techniques – Code of practice for information security controls*

[43] ISO/IEC 33001, *Information technology – Process assessment – Concepts and terminology*

[44] ISO/IEC 40500:2012, *Information technology – W3C Web Content Accessibility Guidelines (WCAG) 2.0*

[45] ISO/IEC 80000 (all parts), *Quantities and units*

[46] ISO/IEC Guide 14:2018, *Products and related services – Information for consumers*

[47] ISO/IEC Guide 37, *Instructions for use of products by consumers*

[48] ISO/IEC Guide 51:2014, *Safety aspects – Guidelines for their inclusion in standards*

[49] ISO/IEC Guide 71, *Guide for addressing accessibility in standards*

[50] ISO/IEC Guide 74, *Graphical symbols – Technical guidelines for the consideration of consumers' needs*

[51] ISO/IEC/IEEE 15288, *Systems and software engineering – System life cycle processes*

[52] ISO/IEC/IEEE 15289:2017, *Systems and software engineering – Content of life-cycle information items (documentation)*

[53] ISO/IEC/IEEE 23026, *Systems and software engineering – Engineering and management of websites for systems, software, and services information*

[54] ISO/IEC/IEEE 26511, *Systems and software engineering – Requirements for managers of information for users of systems, software, and services*

[55] ISO/IEC/IEEE 26513, *Systems and software engineering – Requirements for testers and reviewers of information for users*

[56] ISO/IEC/IEEE 26515:2018, *Systems and software engineering -- Developing information for users in an agile environment*

[57] ISO/IEC/IEEE 26531, *Systems and software engineering – Content management for product life-cycle, user and service management documentation*

[58] ISO TR 16352, *Road vehicles – Ergonomic aspects of in-vehicle presentation for transport information and control systems – Warning systems*

[59] IEC Guide 108, *Guidelines for ensuring the coherency of IEC publications – Application of horizontal standards*
